The Frankenstein syndrome

This book is a philosophically sophisticated and scientifically well-informed discussion of the moral and social issues raised by genetically engineering animals, a powerful technology that has major implications for society. Unlike many other writers on this emotionally charged subject, the author attempts to inform, not inflame, the reader about the real problems society must address in order to manage this technology.

In the first chapter of the book, Professor Rollin examines the claim that genetic engineering, like Dr. Frankenstein's hubristic attempts to create life, is inherently wrong because it is "against nature." Chapter 2 is concerned with the dangers of genetic engineering of animals as embodied in the popular imagination by Frankenstein's rampaging monster. The author proposes a democratic mechanism for public involvement in genetic engineering. Chapter 3 considers a neglected but significant ethical component – the welfare of the created animals. The author explains the new ethic emerging in society for the treatment of animals and shows how genetic engineering must be constrained by that ethic.

Bernard Rollin is a professor of both philosophy, and physiology and biophysics, and he writes from a uniquely well-informed perspective on this topic. His style is nontechnical and anecdotal and will ensure that the book can be used in a wide range of courses on bioethics, biotechnology, veterinary medicine, and public policy. The book should also appeal to a general, nonacademic reader with a serious interest in genetic engineering.

Cambridge Studies in Philosophy and Public Policy

GENERAL EDITOR: Douglas MacLean

The purpose of this series is to publish the most innovative and up-to-date research into the values and concepts that underlie major aspects of public policy. Hitherto most research in this field has been empirical. This series is primarily conceptual and normative; that is, it investigates the structure of arguments and the nature of values relevant to the formation, justification, and criticism of public policy. At the same time it is informed by empirical considerations, addressing specific issues, general policy concerns, and the methods of policy analysis and their applications.

The books in the series are inherently interdisciplinary and include anthologies as well as monographs. They are of particular interest to philosophers, political and social scientists, economists, policy analysts, and those involved in public administration and environmental policy.

Mark Sagoff: *The Economy of the Earth*
Henry Shue (ed.): *Nuclear Deterrence and Moral Restraint*
Judith Lichtenberg (ed.): *Democracy and the Mass Media*
William Galston: *Liberal Purposes*
Elaine Draper: *Risky Business*
R. G. Frey and Christopher W. Morris: *Violence, Terrorism, and Justice*
Douglas Husak: *Drugs and Rights*
Ferdinand Schoeman: *Privacy and Social Freedom*
Dan Brock: *Life and Death*
Paul B. Thompson: *The Ethics of Aid and Trade*
Jeremy Waldron: *Liberal Rights*
Steven Lee: *Morality, Prudence, and Nuclear Weapons*
Robert Goodin: *Utilitarianism as a Public Philosophy*

The Frankenstein syndrome

Ethical and social issues in the genetic engineering of animals

BERNARD E. ROLLIN
COLORADO STATE UNIVERSITY

CAMBRIDGE
UNIVERSITY PRESS

Published by the Press Syndicate of the University of Cambridge
The Pitt Building, Trumpington Street, Cambridge CB2 1RP
40 West 20th Street, New York, NY 10011-4211, USA
10 Stamford Road, Oakleigh, Melbourne 3166, Australia

© Cambridge University Press 1995

First published 1995
Reprinted 1996

Library of Congress Cataloging-in-Publication Data is available

A catalogue record for this book is available from the British Library

ISBN 0-521-47230-X hardback
ISBN 0-521-47807-3 paperback

Transferred to digital printing 2003

To
Linda M. Rollin, Ph.D.,
with thanks for her
unflagging support
and astute criticism

Contents

Contents

Contents

Foreword

"Your manuscript is both good and original," wrote Samuel Johnson (it is said) to a writer seeking his blessing, "but the part which is good is not original and the part which is original is not good."

We can rest easy. Bernard Rollin's ideas, analyses, and syntheses are both good and original, simultaneously. Readers seeking the author's point of view will find it to be that of the well-informed nonscientist, the late-twentieth-century "everyman" who embraces technology only after it is understood in an ethical and social context. We began to understand the genetic code only in the 1950s, and genetic engineering became possible only in the 1970s and 1980s. Thus there has not been time for much "ethical aging" of the issues such engineering raises.

It is in this context that Rollin uses the Frankenstein metaphor as a starting point for his discussion of genetic engineering of animals. No other book on this subject (there aren't many anyway) is as wide-ranging as this one, and none risks putting forth conclusions – in some cases, tentative ones – as this one does. As a result, some scientists will find fault with Rollin's views. But so too will some ethicists and "animal advocates" who will find some of the author's proposals for future genetic engineering surprising.

The more scientists learn about animal sentience, behavior, and self-awareness, the more important are thoughtful analyses of how the application of technology to animals can affect these properties. *The Frankenstein Syndrome* is the place to start.

Franklin M. Loew

*Dean, School of Veterinary Medicine,
Tufts University*

*President, Tufts Biotechnology
Corporation*

*Henry and Lois Foster Professor
of Comparative Medicine,
Tufts University*

Preface

In genetic engineering of animals, as in all areas of applied philosophy, one cannot write intelligently about the ethical issues that arise in the field without first achieving a reasonable grasp of the empirical facts and concepts presuppositional to it. I am thus grateful to the many scientists who have patiently mentored me in the relevant science, and who have in turn been willing to examine that science and its implications through dialectical ethical lenses. I have in fact found most of the people in the field wonderfully open, unthreatened, and kind, and very much concerned about doing the right thing.

Among these scientists who have treated me as a colleague, I must especially single out the following people: Dr. J. Warren Evans, now of Texas A&M, who first challenged me to address the issues growing out of genetic engineering of animals; Kevin O'Conner of the Office of Technology Assessment, who further stimulated my thinking, and Dr. Andrew Rowan, of the Tufts University Veterinary School, who gave me a forum for discussion from which I learned a great deal.

In a class by itself is the debt I owe to my brilliant genetic engineering colleagues at Colorado State University, Drs. Richard Bowen and George Seidel, who, at one time or other, have discussed with me the majority of issues relevant to genetic engineering and whose influence emerges on every page of this book, whether or not they agree with my conclusions. No one could ask for better colleagues or critics. In the same category is my obligation to my wife, Linda Rollin,

Ph.D., whose formidable analytical skills helped keep my arguments tethered to good sense.

I also wish to thank the many colleagues in various fields who helped me with different problems and obstacles: Ken Freeman, Harrison Hughes, Lynne Kesel, Murray Nabors, David Pettus, Mike Rollin, Steve Stack, Ron Williams, and Bruce Wunder. Sandy Woodson and Hollye Gonzales have also earned my thanks for taking my execrable handwriting and turning it into an immaculate typescript.

Finally, I would like to thank Professor David Boonin-Vail of Georgetown University, who, as the reader for Cambridge University Press, generously provided me with extensive constructive criticism, painstakingly detailed. I am grateful for his sympathetic guidance, which far exceeded what one would ordinarily expect.

Introduction

In 1984, I was approached by conference organizers with the request that I give the banquet speech at the first international conference ever held on genetic engineering of animals. Specifically, I was to address the topic of social and moral issues raised by the advent of this new and powerful technology. Flattered, stimulated, challenged, and totally ignorant, I accepted, confident of my ability to rise to the occasion by standing on the shoulders of my predecessors. Unfortunately, a brief visit to the university library shattered my preconceptions – I had no predecessors! My talk, in its published version, would be the first paper ever done on this major topic.[1] Suddenly, I saw my task under a new and harsher light. The buck stopped – and started – with me. Truly an academic's nightmare.

Seeking a purchase on the topic, I solicited dialogue from colleagues in my department. "Genetic engineering of animals," mused one such partner in discussion, "You're talking about the Frankenstein thing." His remark was largely ignored by me at first, as it seemed to me flippant and shallow. It was only later that I realized that he had opened a portal into the issue by forthrightly expressing what in fact rises to most people's minds when genetic engineering is mentioned. A week after our conversation, I was perusing new acquisitions in our university library when I encountered an extraordinary, newly published, five-hundred-page volume entitled *The Frankenstein Catalogue: Being a Comprehensive History of Novels, Translations, Adaptations, Stories, Critical Works, Popular Arti-*

*cles, Series, Fumetti, Verse, Stage Plays, Films, Cartoons, Puppe-
try, Radio and Television Programs, Comics, Satire and Humor,
Spoken and Musical Recordings, Tapes and Sheet Music Featuring
Frankenstein's Monster and/or Descended from Mary Shelley's Novel,* appropriately authored by a man named Glut.[2] The book is
precisely a comprehensive list and brief description of the
works mentioned in the title. After recovering from my initial
amazement that anyone would publish such a book, I was
astonished anew by its content. It in fact lists 2,666 such
works, including 145 editions of Mary Shelley's novel, the
vast majority of which date from the mid–twentieth century.
Putting these data together with my friend's remark, I experienced a flash of insight: Was it possible that the Frankenstein
story was, in some sense, an archetypal myth, metaphor, or
category that expresses deep concerns that trouble the modern mind? Could "the Frankenstein thing" provide a Rosetta
stone for deciphering ethical and social concerns relevant to
genetic engineering of life forms?

In the ensuing months, my hypothesis received succor.
While visiting Australia, I met with an animal researcher
whose field was teratology – the study of birth defects, literally, the study of monsters. He had been extremely surprised
to find that his work with animals had evoked significant
public suspicion, hostility, and protest. "I can't understand
it," he told me. "There was absolutely no pain or suffering
endured by any of the animals. All I can think of is that it must
have been the Frankenstein thing." And in its 1985 cover story
on the fortieth anniversary of the Hiroshima bombing, *Time*
magazine invoked the Frankenstein theme as a major voice in
post–World War II popular culture, indicating that this theme
was society's way of expressing its fear and horror of a science
and technology that had unleashed the atomic bomb.[3]

Those of us who recall the 1950s should be well aware of a
fundamental change during the ensuing decades in social
thought regarding science and technology. At that time, TV
personalities mouthed – with a straight face – the advertise-
ment that we could expect "better things for better living
through chemistry." Sunday supplements were regularly

salted (or peppered) with futuristic articles describing the "New Tomorrow" that science promised – colonization of other worlds, personal "fliers," gold from the sea, robot servants, freedom from disease, unlimited energy from the atom, and so on. Scientists were heroic figures, dragon slayers, in popular culture and its vehicles. The brightest young people aspired to careers in science and engineering.

Four decades later, this naiveté appears laughable to a society cynical and pessimistic about science and technology. Justifiably or not, science and technology have become whipping boys for social ills. (Notoriously, citizens do not distinguish science from technology.) Science is blamed for pollution of air and water, for failing to conquer disease and for iatrogenic disease; for cavalier treatment of human and animal research subjects. Traditional family farms have disappeared with alarming rapidity due to their inability to compete in high technology agriculture. Technology is seen as eroding personal freedom and privacy through computerized record keeping and electronic eavesdropping. Three Mile Island and Chernobyl have perhaps irreversibly damaged public faith in nuclear power. The war on cancer is not won; health care costs are unmanageable; life is often prolonged at great expense to the terminally ill with little regard for life's quality. Computers and video technology are often seen by educators as an alternative to good, human teaching, not as a supplement thereto. And scandals about data falsification, funding misappropriation, and bickering about credit for discoveries in science have helped erode the status of scientists as heroes and demigods. Malpractice suits have done the same for physicians, aided by the ever-increasing tendency of physicians to see themselves as scientists dealing with lawlike phenomena, not healers dealing with individuals. Astronauts have died tragically; space telescopes have been laughably problem ridden. And government surveys objectively document that fewer bright children seek careers in science and that public confidence in science continues to wane.[4]

It is thus no surprise that the Frankenstein story strikes a socially responsive chord, providing us with a way of articulating our fears and doubts about science and technology, a vehicle for packaging and personifying them in a form that allows us to shudder at them, see them vanquished, and go about our business. In and of itself, such personification is benign and even salubrious, purging us, as Aristotle said, of fears we would otherwise suppress. This sort of personified externalization is as old as humanity. On the other hand, when and if the myth becomes reified or transformed into or equated with reality, and thereby conceals nuances, shades, and subtleties of what it represents, the consequences can become socially mischievous.

Something of the latter sort has occurred regarding genetic engineering of animals. The myth has acquired too literal a status in the social mind. People think as follows: Frankenstein created new life with (potentially) hellish consequences; therefore genetic engineering is well represented by "the Frankenstein thing." The myth is either accepted as literal truth or categorically rejected as nonsense, with little thought for the possibilities in between, where the truth surely lies. This dichotomized tendency pervades social thought on genetic engineering and blocks and forestalls any meaningful attempt to place it under meaningful social control or to orchestrate practicable social policies. As an official of the Office of Technology Assessment (OTA; the arm of Congress that advises Congress on new advances in science and technology) informed me when I testified before them: "Congress gets thousands of letters on genetic engineering every week, but they are not helpful. They either tell us to stop all genetic engineering immediately and unequivocally because it violates the laws of God and/or nature; or they demand that we let it forge ahead in an unrestricted way, else we won't be able to compete with the Japanese."

In the end, genetic engineering of animals cannot be stopped – it is too simple and relatively inexpensive to accomplish. Were it banned in the United States, it would simply be moved to a less restrictive (and less regulatory) environment. Not only would we lose patent and incalculable social and economic benefits that can accrue to society (and animals) from its proper use, we would lose control of the technology. On the other hand, the technology needs to be controlled, for a variety of reasons and in a variety of dimensions I will detail in the following chapters.

But as the OTA experience indicates, simplistic acceptance (or rejection) of the Frankenstein story as a way for the social mind to categorize the genetic engineering of animals militates against thinking through the issues that should be considered. And few people have pressed for such consideration; as late as 1986, I was the only person who had written any papers on the moral issues occasioned by the technology, though numerous missives had appeared either extolling its virtues as a ticket to paradise or condemning it unequivocally as a ticket to hell.

As we shall see, what forestalls proper social consideration of scientific and technological advances in biotechnology and other areas is something of a "double whammy." Scientists tend to welcome all new technological and scientific discoveries as positive achievements and to shun discussion of the ethical or social issues they occasion. The public in general tends to be scientifically naive and thus insufficiently knowledgeable to articulate the moral and social issues either. The resultant lacuna is either not filled at all, or it is filled by those with sensationalistic, attention-catching perspectives on the issues. This has been demonstrated over the past two decades with the issue of the care and use of laboratory animals, with the research community insisting that only by having carte blanche with animals could it cure disease, and animal advocates demanding an immediate end to animal research. Fortunately, new (1985) legislation passed between the horns of this dilemma and has moved ahead both the ethic and the treatment of animals.[5] Even more importantly, it has provided guideposts for what is to count as future moral progress in this area and has changed the thinking of many of those who use animals. As a person who was involved both in thinking through the philosophical basis of our moral obligations to animals and in writing what became the federal law implementing that emerging ethic, I have enjoyed firsthand experience in seeing how social moral progress occurs.

A similarly polarizing situation exists currently regarding the genetic engineering of animals, as we have seen. Yet we have hitherto avoided engaging the issues it occasions, both conceptually and pragmatically – this despite the enormous potential power of the technology to change both our lives and the lives of animals, and even the environment. Such a situation is unconscionable and ultimately untenable. We cannot control technology if we do not understand it, and we cannot understand it without a careful discussion of the moral questions to which it gives rise.

Our task, then, in this book is to dissect out these moral issues in the case of genetic engineering of animals and to disambiguate the genuine moral issues from the spurious ones. At the same time, we must also consider the best vehicle for pragmatically dealing with these matters in society.

In order to accomplish this task, we will take the Frankenstein myth as our point of departure. Given the pervasiveness of the myth and given the correlative inevitability of society using the myth as a category for assessing genetic engineering, it is reasonable to unpack its many components and strata in order to lay bare the real issues and dissect out the spurious, knee-jerk ones. Not surprisingly, we will find that while certain aspects of the Frankenstein story do, indeed, admirably express legitimate concerns and reservations, others in fact deflect attention from such concerns and carry the social mind in nonhelpful directions.

CHAPTER 1

"There are certain things
humans were not meant to do"

A GRESHAM'S LAW FOR ETHICS

In the mid–nineteenth century, the British economist Thomas Gresham enunciated the brilliant observation that became known as Gresham's law, that bad money drives good money out of circulation. In other words, if people are faced with the option of paying a debt with either of two currencies of the same face value, they will pay with the one possessing lower intrinsic value. A similar law could be articulated regarding social-ethical questions – especially those associated with science and technology: Bad moral thinking tends to drive good moral thinking out of circulation.

As in the case of money, it is not difficult to see why this should happen. In our society, there is little occasion for reflection on moral questions. This is a fortiori the case regarding those questions associated with science and technology, for one must be scientifically literate before one can engage the ethical questions generated by science.

As I shall shortly discuss, scientists do little to educate the public about moral issues, because most are trained to ignore them. In practice, then, what emerges as "ethical or moral issues" is shaped by the media. The media, as one reporter told me during the Baby Fae case, are "primarily interested in selling papers." What sells papers is what can be packaged in small, provocative bits and what can be dramatically presented as black-and-white extremes. Hence the Baby Fae case was touted as "the Monkey or the Baby"; the issue of the

7

animals used in research is often similarly schematized as "human life or animal life" and so on. It is revealing that during the Baby Fae case, in which the researcher transplanted a baboon heart into an infant, the press did not mention that pig heart valves had been used for years for human heart patients and that, therefore, animal parts had long been "harvested." This case was really not new in terms of the moral questions about animal use it raised. For that matter, the more unusual fact – that the surgery represented unapproved research on a human subject that blatantly violated federal ethical guidelines for such research – garnered few headlines.

And so it goes. When a Food and Drug Administration (FDA) official refused to use toxicological data that the Nazis had gathered in experiments with prisoners, that decision received extensive press coverage. The clear implication was that this was a one-shot oddity, and the issue was laid to rest. The comfortable public perception that the Nazis performed only demented, lurid research into making everyone blond and blue-eyed, or reviving frozen corpses by using naked women to warm them, was untouched. In actual fact, as one might expect from scientists with access to an unlimited pool of human subjects, the Nazi scientists garnered much valuable data. For example, as my colleague Dr. Harry Gorman, a pioneer at the School of Aerospace Medicine, told me, the rapid development of the field of high altitude aerospace medicine, a mainstay of air force and space program activity, would have been a great deal more difficult without the brutal experiments that German Luftwaffe physicians performed on concentration camp inmates. To my knowledge, the tissue of moral questions occasioned by good research arrived at by bad means has never been discussed in the detail it deserves.

In short, the media rarely give ethical questions the detailed analysis they require. But since ethics is by nature provocative, the most provocative formulations of the most provocative questions do get brief and blazing coverage. The public, in its innocence, assumes that the press has covered the important questions and has properly articulated them, and from this comes our social knowledge of scientific or medical ethical issues.

An attendant difficulty with press and video coverage of ethical issues emerging from science, and particularly bioethical issues, is the clear impression imposed on the social mind that all ethical issues are *dilemmas*. Indeed, the mass media tend to equate the phrase "ethical question" with the non-synonymous phrase "ethical dilemma," as if all ethical questions are dilemmas, which of course they are not. A *dilemma*, as the word's Greek etymology implies (di-lemma = two horns), is a situation or question where there are two and only two choices or answers, both of which lead to unsatisfactory conclusions upon which the person faced with the dilemma is impaled. An example of a genuine bioethical or medico-ethical dilemma is provided by the actual case of two people with kidney disease who will die without treatment, while only one kidney dialysis machine is available. Those in charge of treatment are thus faced with a genuine dilemma – if X is treated, Y is untreated and will die; if Y is treated, X is untreated and will die. So dilemmas certainly do exist.

On the other hand, most ethical questions are happily not dilemmas. Indeed, most ethical decisions we make are not even hard choices. For example, many of us may feel the urge to assault, or at least verbally abuse, the innumerable obnoxious people we deal with on a daily basis – taxi drivers, whining students, aggressive panhandlers, heedless drivers, and so on. When we refrain from doing so, we are generally not making a difficult choice but simply suppressing a fleeting impulse, yet we are still making a moral decision. We do in fact make literally hundreds of decisions daily on a regular basis with little deliberation – when we don't cheat on an exam, tell the truth (or tell a little white lie), nip a flirtation in the bud, give or don't give to a charity, punish or fail to punish a transgression by our children, we are making moral or ethical decisions. Many of these are virtually automatic, as when I snarl "Don't bother me!" at Hare Krishna people at airports, but they do reflect principles regarding which we have, at some time, made conscious choices. If challenged on my snarl, I

9

would not say I had yielded to an emotional impulse, but would articulate my reasons for believing that I do not owe even common courtesy to people who use that courtesy as a wedge to badger, hustle, and annoy.

At any rate, the point is that not all or even most ethical questions are dilemmas. But media accounts of ethical issues tend to phrase these issues in ways that suggest a situation is a Scylla and Charybdis one. Part of the reason for this, of course, is that dilemmas generate drama and thus sell papers. But there is a more benign motive as well. Journalists realize that interesting ethical questions have two sides, and in any attempt to achieve fairness, present "the two sides," in their most extreme versions. The implication – or sometimes assertion – which emerges is that there are only two sides, that both have some merit and that, therefore, we have a dilemma when, in fact, logically we do not. In short, various alternatives, intermediary between the extremes, get lost by virtue of the presentation.

The public has a tendency to approach genetic engineering from such a bifurcated perspective, with the only positions that are heard being "stop genetic engineering" or "allow genetic engineering laissez-faire." This, of course, creates high drama and deflects attention away from the via media, which would permit genetic engineering and regulate intelligently those aspects of it that are morally or prudentially problematic. A precedent for this exists in the area of laboratory animals, as I have mentioned and will discuss in detail later; whereas some people demanded that medical research on animals be abolished, and their opposite numbers demanded total autonomy in animal use (an apparent dilemma), still others sought the middle way, which of course prevailed. As any student of American history knows, social change of any significance almost always occurs incrementally, given the diversity of outlooks in our pluralistic democratic society, as a result of the many forces pushing in a variety of directions.

Dualism, or dilemma thinking, is the enemy of compromise and the archenemy of the middle way. As long as people schematize the issue of genetic engineering of animals as "all is permitted" versus "nothing is permitted," rational social progress on the issue is impossible. What is demanded, therefore, is a fair description of the issues, one that separates genuine moral questions from spurious ones, dissects out real concerns from obfuscatory rhetoric, and lays bare truly fundamental areas of concern. Without genuine knowledge of the problems, we cannot generate solutions.

SCIENTIFIC IDEOLOGY AND THE DENIAL OF ETHICS IN SCIENCE

Why have the moral issues relevant to genetic engineering not been brought forward by the scientific community itself? After all, one might reasonably expect that those who work in the field would be most cognizant of the ethical questions it raises. Unfortunately, a variety of factors confound this putatively plausible expectation. In the first place, in general, those working in any field are least likely to think about the conceptual presuppositions and ethical concerns occasioned by the area. Such individuals are so intent on what they do, on working in the area in question, that they rarely can afford the luxury and distancing required to generate thinking about that area. Total immersion in an area tends to be inimical to reflection on that area. Physicians, for example, are generally so busy utilizing or augmenting the armamentarium of contemporary medicine, that they do not (and cannot) stop to examine the questions it occasions, questions of iatrogenesis, escalating economic costs, costs to quality of life associated with heroic efforts at its prolongation, and so on. In the same vein, busy and successful lawyers are not the people to muse upon the ethical questions occasioned by legal practice. As Aristotle pointed out, leisure is the mother of philosophy, and one cannot expect a leisurely perspective on a field from those embroiled in its melee. Plato, too, showed us that bright, busy people in ordinary life resist examination of their fundamental preconceptions.

This generic difficulty, endemic to all fields, is compounded by an ancillary problem unique to science that I have discussed at great length in my book *The Unheeded Cry: Animal Consciousness, Animal Pain and Science*.[1] The failure of the majority of scientists to discriminate moral issues in science is compounded by what I have therein called the common sense of science, or scientific ideology. So important is this notion, that one simply cannot understand the stance of the scientific community on a variety of moral issues in the twentieth century without having grasped its main tenets.

In order to in turn gain ingression into scientific ideology, one must recall some fundamental features of all human activities, particularly those containing any intellectual component. All such activities – art, law, medicine, and science, for example – must rest upon certain assumptions. For example, science assumes that controlled experiments represent a more legitimate approach to understanding nature than anecdotes do; law assumes that people can coherently be held responsible for their actions; medicine assumes a clear distinction between the states of health and disease; art assumes a distinction between those objects crafted by humans that are works of art and those that are not; and so on. As in all of the above examples, the assumptions are prima facie extremely plausible to most people, both in and out of the fields in question. Typically, practitioners in a field do not question the assumptions under which they labor, first and foremost because of the assumptions' plausibility, but also because the practitioners are too busy using them to press the field forward. In addition, there is almost never a context for questioning such assumptions – they work, and as common sense says, "If it ain't broke, don't fix it."

Furthermore, as Aristotle pointed out long ago, all fields of endeavor must rest on certain assumptions that are taken as self-evident. We cannot prove everything, for all proof, as in the paradigmatic case of geometry, rests on extracting conclusions from premises taken for granted. If we could prove our assumptions in a given area, it would be on the basis of other assumptions that were either taken for granted or were proved as consequences of higher order assumptions, which themselves were either taken for granted or extracted from still higher order assumptions, and so forth ad infinitum. Thus, Aristotle concludes, we must either take certain things for granted, or we must be unacceptably spun into an infinite regress.

All of this militates against investigations of assumptions by those who work in a discipline. But that does not, of course, mean that such assumptions cannot be questioned. All assumptions can be questioned, and if they are found wanting or flawed, such questioning can lead to a revolution in the field under examination. This is precisely what occurred when Einstein questioned the traditional assumptions of physics regarding space and time, or when the Supreme Court questioned whether keeping people separate by segregation was conceptually congruent with treating them as equals with those from whom they are separated. Such questioning, which is inherently conceptual in its nature, may come from the field in question, as the case of Einstein eloquently illustrates, or it may come from outside the field. Whenever anyone does inaugurate such a critique of foundational assumptions, though, he or she is functioning as a philosopher, a philosopher of the field in question, for philosophy is precisely the probing and conceptual critique of assumptions in all areas. When most of us think of philosophy, of course, we think of a challenge issued to our most general assumptions about reality, mind, body, right and wrong, truth, beauty, and so on. But philosophy is relevant to esoteric areas as well, as when Russell and Whitehead attempted to clarify the nature of number, or when philosophers of science question the ontological status of such theoretical entities as quanta.

Science obviously rests upon assumptions, both general assumptions common to all of the sciences and specific assumptions made in individual fields and subfields of science. From the early years of the twentieth century on, scientists in all fields have shared a set of assumptions that has almost never been questioned from within science itself. These assumptions attempt to demarcate scientific activity from other activities, such as philosophy and theology, and were articulated in a cluster of related philosophical approaches during the early twentieth century, most particularly, logical positivism, behaviorism, and operationalism. All of these approaches stress that science can deal only with what is verifiable, observable, testable, or subject to experimental manipulation, a theme that one can find even in Newton. Any concepts or claims that cannot be so tested or tethered to observations are disqualified from the purview of scientific activity. Thus, for example, science cannot talk about the soul, or about the "life force" that some nineteenth-century biologists claimed as a key concept of biology, or even about absolute space or time, since none of these can be observed, subjected to experimental testing, or cashed out experimentally.

There is a great deal more to scientific ideology or scientific common sense than this eliminative empiricism, but for our current purposes these additional components can be set aside. Certainly the insistence on not admitting nonverifiable notions into science was salubrious in many cases, for science had, in the nineteenth century, collected much baggage that was mystical or could not be subjected to testing directly or indirectly. Unfortunately, many babies were dumped along with the bathwater. One salient example is provided by the notion of mind, consciousness, and mental states, which were declared to be scientifically unacceptable, a "ghost in the machine," for no one could directly experience anyone else's thoughts. Thus psychology "lost its mind" and restricted its purview to behavior, and prominent psychologists in essence suggested that we do not have thoughts, we only think that we do![2]

As I explored in *The Unheeded Cry*, denial of mind was an unfortunate example of overzealous prosecution of positivistic dicta – as Descartes pointed out, our thoughts and experiences are the primordial stuff of our experience, even our scientific experiences through which we verify our scientific claims! After all, no one can verify anyone else's experiences either – our sense experiences are uniquely ours – and if we do not allow the meaningfulness of talk about what we sense, how can we, in the intersubjective endeavor that is science, gather data? Furthermore, if positivism were really correct, there would be no way to go from individual experiences to the objective world that science allegedly explains – each scientist would be locked into his or her own experiences, unable to correlate them with the subjective experiences of other scientists that are by hypothesis unknowable.

So positivism is untenable and the field of psychology has indeed retreated from behaviorism. But still, its ghost endures, and the vast majority of scientists will still claim that science deals only with "observable facts." For our purposes, though, the epistemological morass created by positivism is not the chief concern. Rather, it is positivism's rejection of talk about values – all values, including moral values – as outside the playing field of science. Since value judgments like "killing is wrong" cannot be verified or falsified by gathering facts or performing experiments, they cannot be considered as part of the fabric of science, given the criteria we discussed above. Thus statements about values are allegedly as irrelevant to science as statements regarding the soul, for they have no empirical tether. At best, they are seen as expressions of emotion, enjoying no claim to truth. This denial of the relevance of value judgments to science is encapsulated in the rubric – ubiquitous among scientists – that science is "value-free."

It is easy to show that this notion, like all components in scientific common sense, is omnipresent in the sciences. From the time that nascent scientists enter their first college course, they are taught positivism and the correlative notion that science is value-free. William Keeton and James Gould, for example, in their widely used freshman biology text, remark that "science cannot make value judgments . . . and cannot make moral judgments."[3] In the same vein, Sylvia Mader, the author of another popular text, asserts that "science does not make ethical or moral decisions."[4] Scientists who worked on the atomic bomb, when queried about the moral implications of their work, responded in essence that ethics is the purview of politicians, their interest was the science – a typical response among those scientists who do weapons research.

15

Few scientific journals, conferences, or courses ever discuss the moral questions engendered by their activity – if they do, it is because society as a whole has become concerned about the issue in question, and researchers must reactively defend themselves. (The response of the medical research community to the issue of research animal use in the early 1980s provides an excellent example of such a reactive posture.) Indeed, the medical research community failed to see or deal with the moral issues involved in the use of human subjects for research until forced to address them by congressional threat of legislation.[5]

It is thus not surprising that leading scientists sometimes say silly things about moral issues in science, if they say anything at all. Donald Kennedy, former FDA commissioner and president of Stanford, remarked, in an attempt to indict critics of animal research, that "antivivisectionism was one of the policies of the Hitler regime."[6] And, in an extraordinarily candid statement directly relevant to the issues of this book, James Wyngaarden, former director of the National Institutes of Health, declared that ethical concern about genetic engineering was misdirected, for "science should not be hampered by moral considerations."[7]

Looking carefully at the sorts of common sense of science statements cited above, one must distinguish two separate claims about moral values in relation to science that appear in scientific ideology and should be kept distinguished from one another. The first sense is fairly plain and can be articulated as follows:

> Claim 1 – The truth of moral statements is unrelated to the truth of scientific statements.

In this sense, the claim is relatively straightforward. It asserts that when one judges the truth of a scientific assertion, one does so independently of appealing to any moral values. That is, when one tests a hypothesis by appeal to experiments or data one has gathered, one is essentially doing something that can be exhaustively described and explicated without reference to moral judgments. Scientists with whom I discuss the issue of morality and science often express Claim 1 when they assert that "science in itself," as opposed to "science as a social phenomenon" or as opposed to "what scientists do," has nothing to do with morality. In other words, the process of confirming or disconfirming a given empirical judgment, which they see as the core of science, involves comparing what one is hypothesizing with what is or is not (factually) the case. Something like Claim 1 is probably what Keeton and Gould or Mader mean to assert in the statements I quoted. In simple terms, a scientific statement like "dinosaurs are extinct" is validated or falsified with no overt or implicit reference to moral judgments.

The second sense, which I shall call Claim 2, is instantiated in the above quotation from Wyngaarden, that "science should not be hampered by moral considerations," and creates the real mischief in scientific ideology. This is the idea that science in the most general sense – not just in the restricted sense of validation of truth claims about the world, but including also the methods of science, the way scientists study phenomena, generally "what scientists do" – is unrelated to morality. We can put this as follows:

> Claim 2 – The truth of moral statements is unrelated to the truth of statements about how the enterprise of testing the truth of scientific judgments should develop.

Contrary to what scientists seem to believe, Claim 2 does not follow from Claim 1. In other words, even if science, in validating truth claims, does not make moral judgments, it does not follow that how scientists go about such validation (for example, whether they use prisoners without informed consent to test hypotheses about drug action) "ought not be hampered by moral judgments."

When the notion that science is ethics-free is interpreted as Claim 2 – that is, as the notion that scientists operate in an ethical void in reference to their methods, choice of how things are studied, choices of what is studied, and so on –

mischief occurs. If ethics is not believed to be scientifically real in the sense of Claim 2, it is no wonder that scientists consistently fail to articulate the ethical issues occasioned by their work, thus allowing that major function to be usurped by the press, as discussed above, or by those with a particular ax to grind. This is bad for society and bad for science and, ultimately, it is baseless, for scientific ideology is clearly wrong. Science cannot be value-free in general or ethics-free in the sense of Claim 2 in particular.

SCIENCE AS ENCOMPASSING VALUES AND ETHICS

The distinction between Claim 1 and Claim 2 will help us understand the sense in which scientific ideology is inaccurate. Given the restricted scope of Claim 1, it is true that morality is not part of science, that is, one can fully understand how science judges the truth of its statements without appeal to the truth of moral statements. On the other hand, Claim 1 cannot legitimately be interpreted as denying the relevance of other nonmoral value statements to the core of science, namely what philosophers call *epistemic value judgments* (i.e., value judgments about what ought to count as knowledge and proper methodology for achieving knowledge). As soon as science indicates a preference for experiment over anecdote, for replicable data over nonreplicable data, for naturalistic explanations over supernaturalistic ones, for quantitative theoretical frameworks over nonquantitative ones, it has made a commitment to certain values, epistemic values, which commitment cannot be confirmed empirically, since this very commitment determines what will count as empirical confirmation! Indeed, the very notion of what will count as a fact, as a legitimate object of investigation, or as data relevant to a given question, rests squarely upon valuational presuppositions.

Consider, for example, the Scientific Revolution, during which the commonsense, sense-experience-based physics and cosmology of Aristotle were replaced by the rationalistic, mathematical, geometrical physics of Galileo and Newton. The discovery of new data or new facts is not what forced the rejection of Aristotelianism – on the contrary, empirical observations all buttressed Aristotle's idea of a world of qualitative differences! What led to the rejection of Aristotelianism was essentially a change in *value* – a discrediting (or disvaluing) of information provided by the senses, as Descartes does so well in his *Meditations*, and a correlative valuing of the rational and mathematically expressible over the empirical, of Plato's philosophy over Aristotle's. This was so nicely expressed in Galileo's claim that, in essence, an omniscient deity would have to be a mathematician and create a mathematical unity underlying apparent diversity.

As I have indicated, Claim 2 about the enterprise of science being unrelated to ethics is the most problematic and mischievous component of scientific ideology, for it is quite clear that, as a social phenomenon and human practice, science cannot be isolated from social morality. The topics that science is permitted or not permitted, encouraged or not encouraged, or funded lavishly or not at all to investigate; the methods science is permitted to employ in order to arrive at knowledge; the types of experiments that are sanctioned and condemned; the way in which human and animal subjects are used in pursuit of knowledge – all of these will depend on the social consensus regarding morality.

The question of invasive use of animal and human subjects provides us with a paradigmatic example of a case where moral judgments are built into the very foundation of current scientific activity. Insofar as we forbear from doing biomedical research on unwanted children or political prisoners, even though they are scientifically a far higher fidelity model for the rest of us than rats or other animals are, we have a moral judgment standing at the very basis of biomedical science. Similarly, if we universally condemn Nazi experiments for failing to operate under principles of informed consent, even if the results obtained are valuable scientifically (which, as I said, was in fact often the case, although press reports tend to suggest that all Nazi research was crackpot), we see scientific advancement checked by moral presuppositions. (As mentioned earlier, a good recent example of this is provided by the FDA's unwillingness to use scientifically valid toxicological data obtained by the Nazis through experiments on concentration camp inmates, even for the benefit of society.) If, as many scientists argue, modern biomedicine is essentially connected to and dependent on the invasive use of animal subjects to function and progress, then there is a series of hidden moral assumptions presupposed here, namely, that the advance of applied knowledge that benefits humans or, for that matter, of pure knowledge with no obvious use, licenses the invasive use of animals, or that the knowledge or control gained through research is worth more than the pain or death of the animals used in that research process.

And if social funding for research is presuppositional to science, for it cannot proceed without resources, then moral notions are again built into science, for such funding is allocated in accordance with social morality and its views of right and wrong. The support of research into disease at the expense of research into gravity waves bespeaks social moral commitments regarding what ought and ought not be done in science. The unacceptability of (and loss of funding to) research dealing with race and intelligence or, indeed, with the measurement of intelligence at all, again provides a clear example, as does the controversy over fetal tissue research.

I have argued elsewhere that the failure of the scientific community to articulate explicitly the moral assumptions upon which the practices of the sciences rest and the moral questions they engender has led to major moral issues in science being ignored. Similarly, such failure to articulate the moral problems occasioned by genetic engineering of animals brings into play the moral Gresham's law I discussed early in this chapter, where questionable ethical questions dominate the social debate, with legitimate ethical questions thrust aside and buried.

"THERE ARE CERTAIN THINGS HUMANS WERE NOT MEANT TO DO": IS GENETIC ENGINEERING INTRINSICALLY WRONG?

This factor is certainly operative regarding genetic engineering of animals, and it is apparent in what is at once the most socially pervasive component of the Frankenstein myth as it applies to the genetic engineering of animals and, at the same time, the least interesting morally. This can be expressed in a variety of ways, but the most memorable would be the classic line from old horror movies – preferably delivered by someone like Maria Ouspenskaya in a generic *Mittel-Europaeisch* accent – that "there are certain things man was not meant to do (or know or explore), and genetic engineering of animals is one of them." This notion, which is indeed older than the Frankenstein story, in fact can be found in the Bible, when God proscribes (albeit not in that accent) the fruit of the Tree of Knowledge. Some theme such as this is often orchestrated in tabloid accounts of genetic engineering – the key notion is that there is something intrinsically wrong with genetic engineering: blurring species, "messing with nature," violating the sanctity of life, "playing God," and so forth. The major point is that the wrongness of the action is not alleged to be a function of pernicious results or negative utility or danger – it is just wrong. This kind of response seems to increase in direct proportion to the dramatic nature of the genetic intervention. If one is putting a human gene into an animal, or vice versa, or genetically creating "monsters," that is, creatures whose phenotype is markedly different from the parent stock, one is much likelier to occasion this response than if one is causing subtle genetic changes, for example, introducing the poll (hornlessness) gene into cattle.

Nonetheless, the pervasiveness of this position cannot be overestimated. An Office of Technology Assessment (OTA) survey showed that 46 percent of the public believes that "we have no business meddling with nature."[8] Similarly, a National Science Foundation survey showed that although members of the public generally oppose restrictions on scientific research, "a notable exception was the opposition to scientists creating new life forms. . . . Almost two-thirds of the public believe that studies in this area should not be pursued."[9] Though survey data are always suspect, these data certainly buttress our claim about the social importance of the first aspect of "the Frankenstein thing."

It is not difficult to find examples of such positions, but it is quite difficult to find arguments to buttress these examples. For the proponents of such a stance, the wrongness appears to be self-evident. But things that are allegedly self-evident in their wrongness are highly suspect – in my lifetime alone, I have heard vehement proclamations trumpeting the inherent wrongness of "racial mixing," of allowing women into male occupations, and of prohibiting corporal punishment of children. So, in examining these positions, one is obliged to try to rationally reconstruct the positions to see if there is any possible coherent argument in their defense, even if such an argument has not been clearly articulated by opponents of genetic engineering. This is the strategy I will adopt in discussing the first version of the Frankenstein thing, which claims that genetic engineering is inherently wrong.

THEOLOGY AND THE ALLEGED INTRINSIC WRONGNESS OF GENETIC ENGINEERING

Probably the most patent source of such a position is theological. When the first genetic manipulations of animals were effected in the early 1980s, for example, when the gene for human growth hormone was inserted into swine, a significant outcry arose from theologians and religious leaders.

The National Council of Churches has declared that genetic engineering of animals does not display proper respect for "the gift of life,"[10] a theme that pervades theological pronouncements on the issue. In the same vein, twenty-four religious leaders issued a pronouncement against animal patenting, employing the following language: "The gift of life from God, in all its forms and species, should not be regarded solely as if it were a chemical product subject to genetic alteration and patentable for economic benefit."[11] Similarly, the same statement asserts that "the combining of human genetic traits with animals . . . raises unique moral, ethical, and theological questions, such as the sanctity of human worth."[12] In none of these pronouncements is any argument, exposition, or explanation provided, so presumably the truth of the position is seen as self-evident.

When one adopts a theological perspective, one can certainly understand the qualms that religious leaders might have about genetic engineering. The Judaeo-Christian tradition has been staunch in its belief that God created living things "each according to its own kind," with the clear implication that species are fixed, immutable, and clearly separated from one another. Nineteenth-century and contemporary opposition by religious factions to Darwin and Darwin's notion of the origin and flux of species illustrates the significance placed upon fixed kinds by religious groups. For humans to meddle with species, to possibly create new species, to blur the lines between species, and, indeed, as Darwin did, to argue that humans and animals are continuous, is to erode the special place of humans and to trade comfortable predictability and order for uncertainty.

More cynically, as one Catholic priest told me in the 1950s, "If humans start creating and radically changing life forms, one ultimate mystery which draws people to religion will disappear. Our hold on people will diminish, they won't need us as badly." In other words, humans will be as gods. The idea expressed by this priest, that God fills in holes where human knowledge is absent and must retreat in the face of the growth of human knowledge, was wonderfully described by Dietrich Bonhoeffer as the notion of "God of the gaps"[13] and represents a primitive form of theology.

Clearly, then, both traditional ideology and rational self-interest militate in favor of conservative church (or synagogue) opposition to genetic engineering. (It should be noted that not all theologians oppose genetic engineering.) What is critical, however, is that such concern, even if well-founded,

23

does not represent a social moral issue. Advances in knowledge and technology that fly in the face of religious tenets may appear morally problematic to adherents to those tenets – many religious people were offended by Newton's physics or Darwin's biology – but that in itself does not create a moral problem for our secular society in general or for its social ethic.

The point is that merely theological concerns do not serve as a basis for asserting in the social ethic that genetic engineering is intrinsically, morally wrong. On the other hand, the fact that a concern is theologically based does not mean that concern has no moral import. To dismiss a concern without examination simply because it may be couched in theological language is to commit a version of the genetic fallacy – confusing the source of an idea with its validity. Indeed, if philosophers like Dewey are correct,[14] putatively religious concerns may well be metaphorical ways of expressing social moral concerns for which no other ready language exists.

REDUCTIONISM AND GENETIC ENGINEERING

Thus one can find in secular critics of genetic engineering such as Jeremy Rifkin, arguments that appear to be couched in religious terms.

> [Genetic engineers] increasingly view life from the vantage point of the chemical composition at the genetic level. From this reductionist perspective, life is merely the aggregate representation of the chemicals that give rise to it and therefore they see no ethical problem whatsoever in transferring one, five, or a hundred genes from one species into the hereditary blueprint of another species. For they truly believe that they are only transferring chemicals coded in the genes and not anything unique to a specific animal. *By this kind of reasoning all of life becomes desacralized* [italics mine]. All of life becomes reduced to a chemical level and becomes available for manipulation.[15]

The notion of *desacralized* is pivotal here. Prima facie, this seems to be a theological notion, yet Rifkin operates from a secular perspective. So one can only guess at what he has in mind outside of a religious context. Presumably he means that one commits some sort of metaphysical transgression when one adopts the stance he deplores in the paragraph quoted above, some sort of secular sin, which has moral import. What this all means is difficult to determine; presumably, he is restating the very point at issue, namely that it is intrinsically wrong to genetically engineer animals. The key concept of desacralization is left undefined or must be thought to be intuitively obvious.

One can deduce from Rifkin's argument that one transgresses against nature when one views life as a bunch of chemicals. Presumably, then, Rifkin is inveighing against the venerable position known as reductionism, which asserts in its epistemological version that all phenomena can be explained by appeal to the laws of physico-chemistry or, in its metaphysical version, that all natural objects are nothing but bundles of chemicals (or molecules, or atoms). Reductionism in one form or another is as old as codified thought: Democritus, for example, the ancient atomist, asserted that only atoms moving about in space were "real"; all else, such as colors and tastes and smells and distinctions between living and nonliving were "by convention" and thence ultimately unreal.[16]

There is no question that a fundamental component of the twentieth-century scientific ideology I discussed earlier is a metaphysical and certainly epistemological commitment to reductionism. Positivism decreed that one did not have adequate explanations of phenomena until one had subsumed these phenomena under the laws of physico-chemistry, which provide us with precision and predictability. Thus most contemporary scientists view molecular biology as more of a real science than organismic or ecological science, because it is expressible in physico-chemical terms while the latter is not. And, as I mentioned, modern medicine has also grown increasingly reductionistic and concerned with the universal and repeatable substratum of disease (medicine as a science), rather than with its unique manifestations in individuals (medicine as an art).

Although reductionism has certainly achieved prominence in contemporary science, it is by no means unquestioned. Classic criticisms of a reductionistic approach were generated by Aristotle and have been reaffirmed by its contemporary critics. Aristotle pointed out that there were in fact four questions or types of explanation one could generate about a given phenomenon: what is it made of, what makes it happen, what is it, and what is its purpose. (He called these the material, efficient, formal, and final causes of a phenomenon, respectively.)[17] As a great defender of common sense, Aristotle pointed out that a good explanation contains all of these components and that reductionism arbitrarily limits itself to the material and efficient causes. One can certainly explain the flight of a bird, for example, in terms of atoms and forces and aerodynamics, but one then loses sight of the fact that what is flying is a bird (its birdness, in his terms) and also of the fact that it is an animal flying for a purpose (e.g., to migrate or breed). This debate between mechanistic reductionism on the one hand, which emphasizes the uniformity of all phenomena and attempts quantitative explanations of all things by appeal to the same set of laws, and teleological materialism on the other, which emphasizes qualitative differences, each explained by laws appropriate to their own domains, continues to rage, and I in fact alluded to it earlier in citing the valuational basis for the Scientific Revolution. In truth, this debate will probably rage forever, for even if we can explain all human behavior, for example, in physico-chemical terms, mechanically, without appeal to intentions or other such concepts, it is an open question whether we should, or whether doing so is adequate.

Thus, for example, I have argued elsewhere, as have many others, that an exclusively reductionistic approach to medicine and disease is highly pernicious. In the first place, it ignores the valuational basis underlying concepts of health and sickness, for value notions and "oughts" are not expressible in the language of physics and chemistry. In other words, when we call a set of physical conditions a disease that ought to be treated, or a state of an organism healthy, that is, what ought to be aimed at, we are making implicit reference to value judgments. It is, after all, a value judgment to say, as contemporary medicine does, that obesity is a disease, rather than a condition conducive to disease; or that alcoholism is a disease, rather than weakness or badness; or that child abusers are sick, rather than evil; or that health is a "complete state of mental, physical, and social well-being," as the World Health Organization definition asserts.[18] Indeed, one legitimate concern about the ability to genetically engineer traits into humans is that it can lead to questionable values informing our concept of health – if everyone can be made 6 feet 2 inches tall, mesomorphic, and blond, and we value this configuration of traits, anyone not exhibiting these traits will be viewed as defective or unhealthy, with a resultant tendency toward loss of diversity in the human population.

In the same vein, reductionistic medicine can be criticized as ignoring the fundamental fact that diseases manifest themselves differently in different individuals, a point dramatically illustrated by Oliver Sacks regarding Parkinson's disease in his *Awakenings*.[19] Similarly, pain experience, even given the same lesion, can vary enormously in many modalities across different humans, across different animals, and across different cultures. This loss of individual perspective – illustrated by physicians' tendencies to talk of "the kidney in Room 407" – can and surely does generate a failure to take account of individual differences, to the detriment of patients.

We may thus surely grant that reductionism may be metaphysically wrong (in attempting to ignore qualitative differences); epistemologically wrong (in allowing for too few types of explanations); even morally wrong, insofar as it leads to pernicious ignoring of real individual differences. But does this allow us to say that genetic engineering of animals is inherently wrong?

I can see no logical basis for such an inference. Even if many genetic engineers are in fact reductionists, it does not follow that they must be. For one can hold a variety of perspectives on the nature of things and still engage in genetic engineering. An Aristotelian could wish to create qualitatively new species, for example – there is in fact a strain in Aristotle, less known than his view of fixed, immutable species, that suggests that creation contains (could contain?) an infinite spectrum of species.[20] One can argue that organisms are more than just bundles of matter – for example, that new properties emerge when we recombine matter – yet still argue both that we can create new organisms simply by manipulating matter and that there is nothing wrong with doing so. By the same token, one could without absurdity be a reductionist and argue that current configurations of matter in motion are "sacred" in Rifkin's sense, or even in a theological sense, and thus should not be tampered with. Such an argument might proceed as follows: God made everything out of the same fundamental stuff according to basic laws but did so in the best possible way, and thus we should preserve, not tamper. Newton might well have so argued.

Recall that we are looking in this discussion for an argument that fleshes out the view that genetic engineering is intrinsically wrong, regardless of the consequences that issue from it. We have thus far looked at Rifkin's implied suggestion that it is wrong because it is tied to reductionism. Even if reductionism is both incorrect and conducive to morally wrong actions, it does not follow that genetic engineering is wrong. To prove this one would need to show that all reductionism is inherently morally wrong, rather than capable of leading or even likely to lead to bad consequences, and that genetic engineering is inherently connected to reductionism, which as I just indicated is surely not the case. I will later discuss the claim that genetic engineering is wrong because it is likely to lead to bad consequences, but that is quite different from the current question – is genetic engineering of animals just morally wrong regardless of consequences?

There is another possible, nonconsequentialist way of fleshing out Rifkin's objection to genetic engineering that, while not explicit in his text, may well be an underlying concern of many people drawn to his unequivocal rejection of genetic engineering. This approach is based neither in saying that genetic engineering is intrinsically wrong (as a theologian might), nor in predicting that genetic engineering is likely to have dangerous or undesirable effects on nature, or to produce more harm than good. Rather such an approach would be based on what is called a virtue-oriented approach to ethics. Such a view, which has recently reemerged in ethical theorizing, harks back to Greek thought, most particularly to Aristotle. In Greek thought, the predominant way of looking at the world was biological, rather than mechanistic. Whereas, since the Renaissance, we have tended to see the world as a machine and to view biology as a subspecies of physics, as reductionism explicitly does, the Greeks reversed those priorities. For them, the world was most like a living thing, and physics was developed by Aristotle by appeal to biological categories. For the Greeks, everything in the world had a nature or a function even as animals and plants do. Thus stones fell to earth because that was their natural place; flames leapt skyward because that was their natural place. Nonliving nature was explained by appeal to categories initially developed to deal with living things.

Ethics was developed in terms of this functionalism. The function of anything – be it a knife or a human being – was that which it alone could do or which it could do best. The virtue of something – its *areté*, or particular excellence – was the trait that enabled it to perform its function. The vice of something, in the same manner, was the trait that prevented it from performing its function. Thus the function of a knife is clearly to cut; its virtue is sharpness, its vice, dullness.

Like anything else in the world, human beings have a nature or function. As Plato shows and Aristotle refines, humans are social, rational, linguistic beings. Their end is to live happy, well-functioning lives in an ordered, well-functioning society. The virtues for human beings are those traits that make such a life possible – justice, courage, wisdom, and so forth. What they have in common is knowledge of the nature of these traits and, ultimately, of the nature of good.

Such an approach can readily be restated in more contemporary terms. Human life remains social; humans remain intellectually based, if not fully rational beings. (Greek optimism about human rationality has been tempered by many modern thinkers who stress the irrational roots of human behavior, and by contemporary events like the Holocaust.) It is still reasonable to inquire into what sorts of things are conducive to well-functioning human social life, though such things are made more complex by the nature of technological society.

Rifkin – or people like Rifkin – may be unequivocally rejecting the perspective on animals and humans that they believe is entailed by genetic engineering, as fundamentally inimical to a good human life. In other words, they may believe that one can only engage in genetic engineering of living things if one views living things as nothing but a bunch of chemicals. Such critics may believe that viewing living things in that way inexorably leads to viewing humans in that way, for we know that humans are animals. Viewing humans in such a manner is in turn not conducive to cultivating the sort of views we find essential to a good society, namely respect for individuals and their dignity, appreciation of their uniqueness, and so on. So the genetic engineering perspective, like the racist perspective, could be said to be fundamentally at odds with the virtues we want to cultivate – and must cultivate – to have a good human society.

This is not a trivial objection. As I mentioned earlier in this chapter, the reductionistic, scientistic approach to the human body dominant in contemporary medicine has clearly had negative results for the treatment of patients. As medicine became more science than art, the individuality of patients was lost, with patients being increasingly perceived as instances of diseases. In perhaps its most pernicious dimension, modern medicine became a matter of preserving life at all costs with quality of life considerations left out of the equation, since there is no room for qualitative notions in reductionistic science. Physicians now cure pneumonia in geriatric patients so that they may die in worse ways, forgetting that pneumonia was once dubbed "the old person's friend." Similarly, as I shall discuss in chapter 3, the gestalt shift on animal agriculture that views it as a matter of industrial management rather than as a matter of husbandry, a shift that emerged some fifty years ago, has had pernicious consequences for the good life both of farm animals and the small farmers who cannot compete with highly capitalized operations.

Thus one can reasonably be alarmed about genetic engineering from the perspective of virtue ethics. But the appropriate response, in my view, is not to attempt to deny the existence of genetic engineering or to attempt to ban it, anymore than the appropriate response to the problems created by modern medicine is to reject it. Genetic engineering, like reductionistic medicine, is a state-of-the-art consequence of the direction science has taken since the Renaissance, and, like it or not, one must deal with it. In any case, we cannot ignore the fact that genetic engineering, as we shall see, can produce a great deal of social benefit; modern medicine has already done so. If properly managed, both can be made harmonious with the good life we seek in society.

The Greeks realized that the key to perpetuating the virtues requisite to a good life and a good society is education. This is why education is perhaps the central theme of Plato's *Republic*, even as it was central to Greek life. *Paideia* was the Greek term for the virtuous education that created complete, virtuous human beings and proper citizens. This is where we as a society have failed regarding the effects of science, particularly reductionistic science, on life and society. We have either ignored the moral effects of such science on life, focusing only on the "advances" flowing from it, or we have tended to reject it in toto, as evil. We do not educate scientists or physicians to be virtuous citizens, we train them in a technocratic way. And, as C. P. Snow long ago pointed out, we do not educate citizens to manage or deal with or even understand the science and technology which ever-increasingly shapes their lives. Indeed, it is only in such a fragmented world that scientists could ever blatantly espouse the ideology that avers that science has nothing to do with ethics (what we have called Claim 2).

We must educate our scientists – be they physicians or genetic engineers – so as to assure ourselves that the moral and social implications of what scientists do are as much a part of their mind-set as are the technical. Equally, we must create a citizenry that understands the science and technology well enough to resist the tendency to leave responsibility for its deployment to the experts. In chapter 2, I will outline a strategy for achieving the latter goal regarding genetic engineering.

We cannot abandon or bury the science and technology we find unintelligible and frightening – though at various times most of us feel that urge. Rifkin notwithstanding, it will not go away or be abandoned. We must manage it. And the only way to do that in a democratic society is to create a common universe of discourse where science and technology, and the moral and social and human problems they can both occasion and help resolve, are regular features of rational, intelligent social discourse. In the case of genetic engineering, we have yet to begin to create such a universe of discourse; we have only barely done better in other areas like medicine. And the only way to guarantee such discourse is through education. Yet, ironically, we live at a time when genuine education has been subordinated to narrow, short-term, expedient, job-oriented training. Reversing that tendency, creating *paideia*, is ultimately the proper response to Rifkin's jeremiads; denying the science and technology of which genetic engineering is the current striking manifestation is simply an ostrich reaction and cannot solve our problems.

GENETIC ENGINEERING AND "SPECIES INTEGRITY"

Let us return to our first attempt to extract a nontheological account of the intrinsic wrongness of genetic engineering of animals from Rifkin's claim quoted above, which I have argued is untenable. The quotation is rhetoric, not argument, and is essentially question begging. Indeed, it is not even coherent in Rifkin's own terms. In the paragraph quoted above, he decries the fact that for genetic engineers, "the important unit of life is no longer the organism, but rather the gene."[21] Yet one sentence earlier, he condemns them for asserting that "there is nothing particularly sacred about the concept of species."[22] A few paragraphs later, he takes these engineers to task as follows:

> What, then, is unique about the human gene pool? Nothing, if you view each species as merely the sum total of the chemicals coded in the individual genes that make it up. It is this radical new concept of life that legitimizes the idea of crossing all species' barriers, and undermines the inviolability of discrete, recognizable species in nature.[23]

The astute reader will have honed in on some flummery in the above or, at the very least, on some fundamental confusion. Is Rifkin bemoaning the fact that genetic engineers do not see the individual organism as the fundamental unit of life, as he sometimes asserts, or the fact that genetic engineers do not respect the species as the fundamental unit of nature? Obviously, both cannot be fundamental.

Indeed, in the space of one page, Rifkin commits himself to two fundamental, ancient, and incompatible philosophical positions – nominalism and realism. Nominalism is the view that the fundamental furniture of the universe consists, as Aristotle said, of unique, particular, discrete, concrete, spatially and temporally located individuals, "this here existent thing," and realism is the view that what is ultimately real are classes, abstract, Platonic entities, which are not located in space and time, species whose reality transcends the spatio-temporal reality of the particular individuals that instantiate the "essence" in question. In the realist view, individual chairs come and go, but the form or essence of "chairness" endures. There are many respectable nominalists and many respectable realists, but one cannot be both, anymore than one can be both a bachelor and married. (As Rifkin tries valiantly but incoherently to be both in the area of metaphysics, so too do many males vis à vis their spousal state.)

In fact, the debate over the ultimate reality of species or individuals in biology, a special case of the nominalism–realism debate, is still extant in biology. Some scientists argue that only individuals are real; species are at best a convenient

artifice for grouping individuals.[24] Others equally vehe-
mently argue that species are the fundamental units of biological reality – species endure and are knowable while individuals come and go.[25] (Interestingly enough, though Aristotle
himself favors the reality of individuals, he concludes that
individuals are unknowable, for we can only separate out
essential features from incidental or accidental ones when we
have a large group to sift through. It is essential to the nature
of humans that humans are rational, but how do we know
whether it is essential or not to the nature of Groucho Marx
that he smoke a cigar?) In fact, little in biology hangs on the
debate – realist biologists must still examine individuals, and
nominalist biologists still acknowledge species. But these different metaphysical positions do generate very different
stances on a variety of bioethical issues, including the moral
status of animals, environmental ethics, and genetic engineering, so the issue must be addressed.

As we shall see, one encounters the appeal to the inviolability of species in a variety of quarters. The illegitimacy
of crossing species barriers, the inviolability of species, comes
up again and again in Rifkin's writing and in the pronouncements of those others who see genetic engineering of animals
as intrinsically wrong. So the question arises, what sense can
be made of the notion of species inviolability? Our paradigm
case for the wrongness of "violating" something in a morally
relevant way comes from our consensus ethic for the treatment of human beings. We "violate" people by causing them
physical or psychological harm, frightening them, humiliat-
ing them, stealing or destroying their possessions, thwarting
their freedom, not allowing them to express their natures,
violently intruding into their physical or emotional lives, and
so forth. And this is considered immoral or wrong because
what we do to humans matters to *them*; thwarting their interests and desires and needs makes them suffer. We will later
see how society is plausibly extending this notion to animals,
where we have good reason to believe that what we do to
them matters to them. On the other hand, we cannot violate –
or hurt – a rock, or a car, or a tree, except metaphorically, since

what we do to nonsentient entities doesn't matter to them. To be sure, we can "violate" a person's house or car, as when we force our way into them and do damage, but we are harming the person who owns the house or car, or those who could use them, not the objects in a morally relevant sense. We cannot wrong the car.

Similarly, can I harm or violate a species? Humans can decimate a species, like the buffalo, or destroy it completely, as in the case of the passenger pigeon, but have we done harm to the species? In my view, we have harmed in a morally relevant way animals who comprise the species; or perhaps the humans who depend on those animals or plants or who admire them aesthetically; or the humans who care about biodiversity; or else the animals that depend on the members of the vanished species. But we have not harmed the species, because species are not sentient; only the members of some species are. A species, if it has any existence above and beyond the members thereof, exists as an abstract Platonic entity like a set or a number. We do not harm the number one by destroying all the things in the universe it can denote; for a Platonist, it continues to endure in splendid isolation. By the same token, the class (or species) of buffalo does not care whether it has few or many members, though the buffalos presumably do.

To be sure, all right-thinking people are or ought to be greatly disturbed when a species, containing sentient members or not, is threatened with the extinction of its members. Why? Because dodoes, or Siberian tigers, or an endangered moss will never pass this way again. Because they may well be inextricably linked with the survival of other sets of individuals we or others care about or, indeed, with our own survival. Because they are beautiful, and we want our children and ourselves to be able to experience them. But we can also care about the defacing of the Mona Lisa or the erosion of the Sistine Chapel without granting them moral standing in themselves, but strictly on the basis of their relations with sentient beings. As far as we are concerned, it may be far worse to kill the last ten Siberian tigers than ten other Siberian tigers when there are many of them. (This is, of course, the basis of the conservation ethic.) But as far as the tigers are concerned, they don't know or care whether they are the last, and thus it is equally wrong from a tiger's perspective (if it is wrong) to kill any ten or the last ten. And as I said earlier, none of it matters to the species at all, for things do not matter to species, and, if one is a realist, species continue to exist whether or not their members do!

The mistaken tendency to identify species as sacred things has many sources. First and foremost, as we have already seen, is the biblical notion that God created kinds. And if we annihilate – or otherwise meddle with – kinds, we are meddling with God, to whom, as the Bible indicates, our actions matter a great deal. Second, the dominant version of Aristotelianism that has pervaded Western culture also stresses the fixed and immutable nature of things. Aristotle's reasons for believing in fixed and immutable species seem to come from two fundamental sources. In the first place, as a commonsense philosopher, Aristotle points out that if species changed, we would surely see cases of such transmutation, yet this has never been noted by human beings. (Aristotle and most of his contemporaries did not view fossils as evidence of vanished beings – they rather saw them as kinds in themselves. Thus fossil fish were stone fish, not traces of bygone fish.) Second, again operating from a commonsense basis, Aristotle points out that if natural kinds changed, knowledge would be impossible, since knowledge entails stability, and nothing is plainer than the fact that humans are by nature "knowing creatures." Thus both common sense and Aristotle's articulation thereof loom large in Western culture's tendency to see natural kinds as fixed.

Yet another factor militating in favor of viewing species as fundamental units of reality has come from the scientific community – significant portions of the biological scientific community have argued that species are in some sense "more real" than other units of classification. In other words, in this view, species concepts are more than arbitrary, Dewey-decimal system sorts of classificatory schemes, but rather somehow accurately subtend the way things are "in the world." Though biological taxonomists have differed as to the way in which species are characterized – some, for example, have located the nature of species in genetic similarity, others in genealogical or evolutionary history – both approaches would agree that species locutions are closer to reality than the other classificatory concepts commonly employed, which are more inclusive than species. These are, as any biology book recounts, in decreasing order of generality, kingdom, phylum, class, order, family, genus, species. Thus, the domestic dog belongs to the animal kingdom, phylum of chordates, class of mammals, order of carnivores, family of canids, genus of canis, and species *canis familiaris*.

Michael Ruse, a philosopher of biology, has succinctly summarized the view of biologists regarding the reality of species as follows:

> There seems to be common agreement amongst biologists . . . that there is something rather special about the biological species. . . . Somehow, groups which are biological species are felt in some sense to be "real" in a way that other groups are not felt to be. . . . Almost without exception, evolutionary taxonomists are adamant in their contention that it is biological species alone which are real.[26]

Interestingly enough, the root of this idea is probably in part common sense. Ernst Mayr, the classic defender of the reality of species, argues that the reality of species is apparent in the universal way in which humans delineate groups of animals and plants as separate from one another:

> The primitive Papuan of the mountains of New Guinea recognizes as species exactly the same natural units that are called species by the museum ornithologist. The arrangement of organic life into well-defined units is universal, and it is this striking discontinuity between local populations which impressed the naturalists Ray and Linnaeus and led to the development of the species concept. There can be no argument as to the objective reality of the gaps between local species in sexually reproducing organisms.[27]

This sort of approach is indeed reflected in the standard biology textbook account of species; for example, one recent text asserts that a species is "a genetically distinctive group of natural populations . . . that share a common gene pool and that are reproductively isolated from all other such groups."[28] Another text asserts that a species is a group whose "members can breed successfully with each other but not with organisms of another species."[29] (We can charitably forgive the patent circularity in the above definition, as it does not interfere with our point.) Still other definitions add that members of a species must be able to both successfully interbreed and produce fertile offspring – this qualification makes clear that animals like horses and donkeys are not in the same species, as they can and do successfully reproduce, but they do not produce fertile offspring (mules being infertile). On the other hand, the definition is still not adequate, as there are in fact groups we call separate species that breed and produce fertile offspring – dogs and wolves provide a familiar example, as do dogs and coyotes; lions and tigers can also be bred to produce fertile offspring. By the same token, Chihuahuas and Great Danes are considered members of the same species, yet they are clearly unable to breed without human help and thus do not meet the definition of species.

It is widely held in biology that there is, in fact, no adequate definition of species that is not subject to refutation by counterexample. The notion of species thus appears to be a fuzzy concept, one that is not precise, or definable, but that captures our intuitive tendencies about grouping what we find in the world. And given the evolutionary paradigm, species are going to have very fuzzy boundaries, as new species emerge from old species by new selection pressures that favor changes that have emerged by chance.

Thus modern biology does not accept the notion of species as fixed, immutable kinds, a notion biblically enunciated and articulated in Aristotle. In post-Darwinian biology, nature is forever experimenting with modifications of extant species, most of which modifications will prove to be deleterious, but a small number of which will contribute to the incremental (or, according to some theorists, dramatic) changes that eventually produce new species. One can therefore argue that, if species are not fixed, as far as nature is concerned, and humans are part of nature, there is no reason to say that humans cannot produce new species – we have certainly done so unwittingly in countless cases by radically altering environments. Indeed, we have caused the extinction of countless species in the course of human civilization, and, as environmentally concerned scientists point out, are doing so now at an unprecedented rate. In fact, it may well be that concerns about annihilation of species at human hands spills over into concerns about changing species – it is a small step, psychologically, though logically an indefensible one, to go from the claim that species ought not to become extinct due to human actions to the claim that species ought not to change at human hands.

We have certainly drastically altered species in an intentional way, ever since we have been able to do so. And we have in fact created new plant species in abundance through hybridization – the tangelo and some types of orchid are mundane examples of such genetic manipulation. Indeed, it is estimated that 70 percent of grasses and 40 percent of flowering plants represent new species created by humans through hybridization, cultivation, preferential propagations, and other means of artificial selection.[30] Few people feel that such breaching of species barriers is intrinsically morally problematic. On the other hand, though we have produced mules, beefalos, tiglons, ligers, coydogs, and other sterile and non-sterile hybrids in the animal kingdom, we have never been able to produce a new species, in the sense of animals so far removed from the parent stock that they are reproductively isolated from it but still fertile. Nor have we yet done so via genetic engineering; nevertheless, this is clearly what drives concern of the sort enunciated by Rifkin, for it is surely in principle possible, though not practically possible given existing technology.

So it appears that, for the immediate future at least, genetic modification of animals will not breach the species barrier but will accelerate the sorts of modifications that we have hitherto effected using artificial selection. Breeding for phenotypically evident traits such as greater size, quicker growth, and disease resistance, does not involve creating new species and thus does not differ in kind from what we have always done. It does, however, differ in the rapidity with which changes can be introduced into a genome; as one of my colleagues put it, current techniques of genetic engineering allow us to produce changes "in the fast lane." This in turn means we do not go through a long period of trial and error over many generations during which time we can detect untoward consequences of our engineering that can cause harm to the animal or to humans or to other animals. But this in turn does not provide a reason for believing that genetic engineering is intrinsically wrong, while traditional manipulation of genomes through breeding is not. Rather, it tells us that genetic engineering may occasion undesirable consequences. Perhaps this risk is so great as to militate against manipulating animals in the fast lane – I will address the dangers of genetic engineering later – but then the wrongness of genetic engineering is not intrinsic, but consequential. And we are here attempting to unpack the possible arguments for its allegedly intrinsic wrongness.

NATURE, CONVENTION, AND GENETIC ENGINEERING

It may well be that behind the opposition to genetic engineering of animals as intrinsically wrong – the first version of "the Frankenstein thing" – lies an age-old metaphysical dualism that has colored Western perception of reality at least from the time of the pre-Socratics, the dualism of nature and convention, *physis* and *nomos*, nature and culture. This dichotomy presents a radical separation between what is true by nature and what is true by human construction. We have already questioned this distinction as drawn by Democritus; it persists throughout the history of Western thought. I have, in my own work, traced the dualism as it applies to natural and conventional signification or meaning from antiquity through the present, and I have found it to be more confusing than enlightening.[31] Commonsensically, if humans are a part of nature, and subject to its laws, they can hardly be radically discontinuous with nature.

The assumption that all phenomena must fall into either that which is natural or that which is conventional has conditioned our thinking and, I believe, distorted it. I alluded earlier to the controversy over the reality of species versus individuals. The debate has, as we saw, been largely depicted as raging between realists, who believe that species are real, and nominalists, who believe that species designations are arbitrary or conventional. In my view, recognizing that all phenomena need not be clearly either natural or conventional, and allowing for elements of both, provides us with a third way of talking about species. As I have argued elsewhere, the way we classify the world is in part based on the theories we embrace to help us do the classification.[32] Our concepts are filters through which we sift our experience, and thus which concepts we use will surely have an effect on the world we experience. To take a powerful example, psychiatrist Robert Jay Lifton has shown that German medicine and medical training during the Nazi era was permeated with the idea that the state was a living body. As such, it could host pathogens injurious to its health. Non-Germans, such as Jews, Gypsies, and retarded people, were seen as pathogenic to the body politic, and their removal analogous to ridding a person of germs, viruses, or parasites.[33] Seeing through these glasses, otherwise morally sensitive physicians could bring themselves to euthanize defective children – an act that, unfiltered through that ideology, would otherwise be monstrous. Note that I am not exonerating the Nazis, or saying that their conceptual scheme was justifiable, or asserting a relativism among conceptual schemes. I am only arguing that different conceptual schemes do exist and do shape our experience. Conceptual schemes can and do color the form that science takes – I have already discussed the view that the Aristotelian metaphysical commitment to a world of qualitative differences made for a science very different from the one stressed by Descartes, Galileo, and Newton, which placed major value upon quantitative dimensions of phenomena.

In any event, my point is that the nature-convention dichotomy is pervasive in Western thought and has buttressed the cleavage between those who think species are real and those who think they are constructs. My third way suggests that how we classify is an admixture of theory and "fact," nature and convention. What species or kinds we find in nature depends on the theories we hold. If we are committed to an evolutionary/genetic approach to biology, certain empirical tests will emerge to tell us what things are or are not related – DNA sequencing, for example. On the other hand, suppose our culture was primarily interested in classifying according to ecological or behavioral concerns (or values). In that case, we would not look to DNA sequencing for evidence as to classification, but perhaps to similarity of function or niche filling. We could still have empirical tests to decide what was to be grouped with what; they just would not be the same tests driven by an evolutionary-theoretical approach.

Similarly, the nature-convention split in fact plays into scientific ideology and has mischievous effects on science and medicine. As we saw, science tends to eschew values and believes itself to deal only with facts. Facts are, according to scientific ideology, given by nature; values are provided by human conventional construction. Values, as we saw earlier, are believed by scientists to be subjective, arbitrary, culturally variable, humanly constructed – in short, to meet many if not all of the classic criteria for what is conventional, not natural. Thus, insofar as the duty of science is to deal with and lay bare the verities of nature according to scientific ideology, this mission would militate in favor of distancing science from attempting to engage valuational matters. Not only does this help blind science to ethical questions, as I discussed earlier, it also tends to oversimplify certain phenomena that science studies.

A paradigm case of the mischief engendered by this mindset arises in medicine. Physicians are convinced that a judgment that something is diseased or sick is as much a matter of fact as is the judgment that the organism is bigger or smaller than a breadbox. Diseases are repeatable entities to be scientifically discovered – physicians are scientists. This scientistic stance has been repeatedly noted in its nonsubtle manifestations; as I discussed earlier, anyone who has been in a hospital is aware of the tendency of physicians to see patients as instances of a disease rather than as unique individuals – science after all deals with the repeatable and lawlike aspects of things, not with individuals qua individuals. (As mentioned earlier, Aristotle recognized that science could not know particulars.) This tendency to remove individuality is a chronic complaint of patients – it is demeaning to be treated as an instance of something. Indeed, it is less often noticed that this tendency is medically pernicious as well. When it comes to dispensing pain medication, for example, it has been shown that pain tolerance thresholds (i.e., the maximum pain a person can tolerate) differ dramatically across individuals and that thresholds can be modulated by a variety of factors,[34] not the least of which is surely rapport with the physician, or the sense that the physician cares about the patient's pain. Among physicians, only Oliver Sacks, in *Awakenings*,[35] as I mentioned earlier, has stressed the extraordinary degree to which a disease varies with the individual, in all of his or her complexity, in whom that disease is manifested or instantiated.

This much ordinary common sense (but not medical common sense) recognizes. The more subtle sense in which scientism in its emphasis on fact versus value and nature versus convention – with only the former term in each pair entering into the medical situation – misses the mark is in its understanding of the very nature of disease. For the concept of a disease, of a physical (or mental) condition in need of fixing, is inextricably bound up with valuational presuppositions. Consider the obvious fact that the concept of disease is a concept that, like good and bad, light and dark, acquires its meaning by contrast with its complement, in this case, the concept of health. One cannot have a concept of disease without at

43

least implicit reference to the concept of health (= okay and not in need of fixing). Yet the concept of health clearly makes tacit or explicit reference to an ideal for the person or other organism; a healthy person is one who is functioning as we believe people should. This ideal is clearly valuational; most of us do not feel that people are healthy if they are in constant pain, even though they can eat, sleep, reproduce, and so forth. That is because our ideal for a human life is really an ideal for a good human life – in all of its complexity.

Health is not merely what is statistically normal in a population (statistical normalcy can entail being diseased); nor is it purely a biological matter. The World Health Organization captures this idea in its famous definition of health as a complete state of mental, physical, and social well-being. In other words, health is not just of the body. Indeed, the valuational dimension is both explicit and not well defined, for what is "well-being" save a value notion to be made explicit in a sociocultural context?

Heedless of this point, and wedded to the notion that disease is discovered by reference to facts, not in part decided by reference to values, physicians make decisions that they think are discoveries. I alluded to this earlier in my discussion of value-free science. When physicians announce that obesity is the number one disease in the United States, and this "discovery" makes the cover of *Time* magazine, few people, physicians or otherwise, analyze the deep structure of that statement. Are fat people really sick people? Why? Presumably the physicians who make this claim are thinking of something like this: Fat people tend to get sick more often – flat feet, strokes, bad backs, heart conditions. But, one might say, is something that makes you sick itself a sickness? Boxing may lead to sinus problems and Parkinson's disease – that does not make it in itself a disease. Not all or even many things that cause disease are diseases.[36]

Perhaps the physicians are thinking that obesity shortens life, as actuarial tables indicate, and that is why it should be considered a disease. In addition to being vulnerable to the previous objection, this claim raises a more subtle problem.

44

Even if obesity does shorten life, does it follow that it ought to be corrected? Physicians, as is well-known, see their mission (their primary value) as preserving life. Others, however, may value quality over quantity of life. Thus, even if I am informed – nay, guaranteed – that I will live 3.2 months longer if I drop forty-five pounds, it is perfectly reasonable for me to say that I would rather live 3.2 months less and continue to pig out. In other words, to define obesity as a disease is to presuppose a highly debatable valuational judgment.

Similar arguments can be advanced vis à vis alcoholism or gambling or child abuse as diseases. The fact that there may be (or are) physiological mechanisms in some people predisposing them to addiction does not in and of itself license the assertion that alcoholics (or gamblers) are sick. There are presumably physiological mechanisms underlying all human actions – flying off the handle, for example. Shall physicians then appropriate the management of temperament as their purview? (They have, in fact.) More to the point, shall we view people quick to anger as diseased – Doberman's Syndrome?

Perhaps. Perhaps people would be happier if the categories of badness and weakness were replaced with medical categories. Physicians often argue that when alcoholism or gambling is viewed as sickness, that is, something that happens to you that you can't help, rather than as something wrong that you do, the alcoholic or gambler is more likely to seek help, knowing he or she will not be blamed. I, personally, am not ready to abandon moral categories for medical ones, as some psychiatrists have suggested.[37] And, as Kant said, we must act as if we are free and responsible for our actions, whatever the ultimate metaphysical status of freedom and determinism.[38] I do not believe that one is compelled to drink by one's physiological substratum, though one may be more tempted than another with a different substratum.

Be that as it may, the key point is that physicians are not discovering in nature that obesity or alcoholism are diseases, though they think they are. They are, in fact, promulgating values as facts and using their authority as experts in medicine

45

to insulate their value judgments from social debate. This occurs because they do not see that facts and values blend here, that nature and convention intertwine. They are not ill intentioned, but they are muddled, as is society in general. And to rectify this, we must discuss, in a democratic fashion, which values will underlie what we count as health and disease, not simply accept value judgments from authorities who are not even cognizant of their existence, let alone conceptually prepared to defend them. At the very least, if we cannot engender a social consensus, we should articulate these for ourselves. Here again, we see the need for education, *paideia*, which allows these divergent values to be engaged in one universe of discourse.

Thus the traditional, rigid dichotomy between nature and convention, which is ubiquitous in Western thought, likely buttresses the idea that genetic engineering of animals is intrinsically wrong and may lie behind the notion of the inviolability of species integrity. If the realm of nature is given, and distinct from the realm of human artifice, it represents a metaphysical sin, as it were, to introduce artifice into nature. The response to this, of course, is that humans do intentionally influence nature in myriad ways, from the domestication and breeding of domestic animals to the cultivation of plants, the diverting of rivers, and the leveling of mountains. One may indeed affirm that it is natural to humans to effect conventional changes in nature, once again pointing to our inability to clearly separate these categories. I am not, of course, denying that there are paradigm cases of things that occur by nature and things that occur by convention; the fact that water flows downhill is clearly natural, the fact that the phoneme "Hee" means "he" in English and "she" in Hebrew is clearly conventional. What I am rather suggesting is that there is a spectrum between these extremes and shadings and blending of both notions in many phenomena, not a metaphysically rigid dualism.

Further, I have argued that the tendency to bifurcate phenomena into either that which exists or is true by nature or that which is true or real by convention potentiates the scientific ideology that excludes values from the domain of science. Scientists then have additional reasons for eschewing talk of the value issues underlying genetic engineering of animals, at the same time as society as a whole feels strongly but inchoately that major value issues, particularly bioethical issues, are in fact being raised by genetic engineering. This vacuum is quickly filled by the sorts of insubstantial claims we have been considering. Ironically, therefore, the nature-convention dichotomy is perniciously operative twice in the issue of genetic engineering of animals. First, it helps scientists to ignore ethical issues in genetic engineering; second, it at the same time engenders metaphysically biased "ethical" pseudoissues in the general population.

ENVIRONMENTAL PHILOSOPHY AND GENETIC ENGINEERING

Also connected, in an elliptical but powerfully seductive way, to the mind-set that sees genetic engineering of animals as intrinsically wrong is another mixture of prudence, ethics, and metaphysics, which has lately come to figure prominently but confusedly in Western thought. This is *environmental philosophy*, an odd amalgam of sound, genuine concern with fouling our own nest (prudence); overblown ethical system building designed to provide a new ethic where none is needed; and wild metaphysical pontification designed to ground the unneeded ethic. From this potpourri inexorably comes a mind-set that, when tied to the nature-convention dualism, sees genetic engineering of animals as intrinsically wrong, for nature is portrayed in this theory as intrinsically valuable – indeed, more valuable than individual humans – and human meddling with valuable natural entities, species, for example, as morally wrong.

Now there is no question that ethical thinking relevant to how we approach our environment has been sorely lacking in human history, at least in technological societies. So-called primitive societies have long realized that respect for the natural world was a necessary condition for survival, and such respect was mirrored in their religions, mythologies, and social organization. A complex example of such a culture may be found in Australian aboriginal society, highly stable in its fundamental tenets for millennia. Such harmonious stances were clearly born of necessity; the price of a cavalier stance toward nature was a precipitous diminishing of one's chances of survival.

With the rise of technology, however, there came an apparent ability to tame or harness nature. When this was coupled with the strong Christian tendency to deny any sacred status to nature (in part to distance itself from autochthonous religions) and to view nature as provided by God for human use, the stage was set for environmental despoliation. With the advent of the Industrial Revolution's "dark satanic mills" came fouling of air and water and soil on an ever-increasing scale. Species of plant and animal were driven to near-extinction and extinction to provide for all manner of human desires and to create wealth from "natural resources." As technology grew exponentially, and urbanization increased, people grew less sophisticated in their understanding of human dependence on nature. With this came an attitude that humans were improving the natural world, adding value to a limitless supply of raw material. And with this came also a new hubris, that we could improve on nature endlessly at little or no cost, a stance that well characterizes our ever-increasing dependence on agri-chemicals, and that nature was not only an endless source of bounty but an endless and forgiving receptacle for the waste products of high technology. Indeed, I distinctly recall my mind-set and that of my peers in the early 1960s – an unlimited confidence in science and technology to positively dominate and control and overcome nature at little or no cost, a mind-set that I alluded to earlier in this chapter.

In my own case, which I suspect is rather typical, I was "shocked out of my dogmatic slumber," in Kant's felicitous phrase, by a series of related incidents. The first occurred when, in 1962, I was caught in the rain one beautiful spring day in New York City and found that wherever my bare skin

48

had been exposed to the shower, the skin was burning, itching, and inflamed. Upon rushing to my physician, I found that I had experienced an extreme reaction to acid rain, resulting from air pollution. Three years later, the same polluted air was driving me to the emergency room of St. Luke's Hospital on a regular basis, often three or more times in a week, the victim of chronic asthma. By 1967, I was quite conscious of issues of air quality and was appalled to find, when touring remote portions of northern Canada, that the people living there, under subsistence conditions, were afflicted with extreme air pollution from U.S. Midwestern factories, while at the same time reaping none of the benefits emerging from that industry.

Thus when the Phi Beta Kappa chapter at the City College of New York surveyed its members in the mid-1960s on what they considered the most significant problems facing American society, I was one of the only people responding to put environmental concerns at the top of the list. By the early 1970s, this had changed substantially. Today the vast majority of Americans consider themselves environmentalists, and even kindergarten children worry about Mother Earth, recycling, and the devastation of the rain forests; and terms like *biodiversity* are household words.

This is, of course, all to the good. For reasons that are not clear, however, a group of environmentally concerned philosophers have felt compelled to generate a radically "new ethic" for the environment and have argued that natural objects (concrete and abstract) – ecosystems, rivers, species, and nature itself – possess intrinsic or inherent moral value and are direct objects of moral concern to which we have moral obligations. Indeed, for many of these theorists, these entities have higher value and more inherent worth than "mere individuals," human or animal. By arguing this case, such theorists hope to introduce a solid grounding for environmental concern. This, in turn, tends to foster a "nature is perfect as it is" attitude, which naturally fuels aversion to changing species or making other modifications by genetic engineering and, indeed, suggests that this is morally wrong.

I am not here asserting that a commitment to intrinsic value in nature logically entails the rejection of genetic engineering as intrinsically wrong. I am rather saying something weaker, namely, that many people who would be drawn to a position that ascribes intrinsic value to nature would tend to view human meddling with species (which allegedly possess intrinsic value) as morally problematic, indeed, as intrinsically wrong. The connection between the two attitudes is similar to the one that we have seen obtains between conservative theological positions and the view that genetic engineering is intrinsically wrong. It is thus worth our while, I believe, to look in more detail at the arguments underlying the position of those who adopt environmental philosophy in the sense outlined above.

Unfortunately, it is difficult to get a purchase on their arguments, though, fortunately, one can generate an environmental ethic without recourse to what a colleague calls "spooky" commitments to the inherent value of rocks and trees. In the first place, we need to recall – and this is highly relevant to the alleged intrinsic wrongness of genetic engineering – as Plato stressed, that those wishing to criticize or advance ethical positions cannot teach, they can only remind. That is, teaching ethics is not like teaching state capitals or chemistry – I, as a moralist, cannot walk in and give you a list, ex nihilo, of what is right and wrong and expect you to accept it. The natural response to such a move is "Says who?" or "Why should I accept what you say?" To be plausible, I must extract (or have you extract) my conclusions from ethical premises you already accept as true, but which extraction you have hitherto not done. The moral philosopher, says Socrates, is a midwife.

I shall later provide numerous examples of such ethical midwifery. Suffice it to say that when one is working on sensitive ethical issues, the deployment of that technique is a necessary prerequisite to achieving any measure of success, or often, to establishing any communication at all. For example, when I first began to lecture to groups of farmers and ranchers on animal welfare/animal rights issues, the audience hostility was palpable, and often overt. In order to defuse it, I would ask the audience to answer two questions in order to actuate the process of "reminding." "First of all," I would inquire, "Do you guys believe in right and wrong?" "Hell, yes!" was the invariable rejoinder. "Second," I would continue, "Would you guys do anything at all to an animal to increase profit and productivity? For example, would you torture a cow's eyes with hot needles if it increased milk production?" "Of course not!" they would reply. "Good," I would say. "Then we are just arguing about price." Invariably we could then begin a constructive dialogue in a hostility-free context.

In my last chapter, I will show how the social ethic for the treatment of animals is growing and deepening, precisely through a process of delivering what is already implicit in our ethic for humans. I will argue that if there is any sense to the notion of intrinsic or inherent value, something we ascribe to humans, it grows out of the fact that humans are capable of valuing themselves; in other words, as conscious or sentient beings, what we do to them matters to them. They are capable of happiness or pleasure or satisfaction on the one hand, and unhappiness or pain or dissatisfaction or frustration on the other. In other words, the ability to value (or disvalue) what happens to them, which humans have by virtue of being conscious, makes value inherent in them. Since (many) animals have this same capacity of sentience, the notion of inherent value can (and indeed must) reasonably be extended to them.

Rocks, species, ecosystems, forests, wilderness areas, and so forth do not have this sort of capability. Thus we cannot extract their intrinsic worth from our consensus social ethic. To my knowledge, none of the environmental ethicists we are discussing have argued that these entities are conscious and can be harmed in ways that matter to them.

Wherein, then, do these theorists locate the intrinsic moral value of these entities that allegedly grounds our moral obli-gations to them? This is not at all clear in their writings. If we look, for example, at the discussions of intrinsic value by Holmes Rolston, perhaps the leading environmental ethicist, we find a number of different defenses of such value in nature.

Throughout his discussion of wilderness, nature, species,

51

natural objects, and so on, Rolston is insistent that such objects possess intrinsic value, not merely use value for humans,[39] and that this value is really there in nature, independently of whether humans realize or experience it.[40] His arguments in defense of this claim take various forms. One common argument proceeds as follows:

Argument 1: Because we apprehend beauty or other aesthetic experiences when we interact with nature, nature has inherent value. Such claims appear throughout Rolston's writings. For example, in his essay "Can and Ought We Follow Nature," he makes the following statements:

> The rural environment is an end in itself as well as an instrument for the support of the city. It has beauty surpassing its utility.[41]

> We need wild nature in much the same way that we need the other things in life that we appreciate for their intrinsic rather than their instrumental worth, somewhat like we need music or art, philosophy or religion, literature or drama. But these are human activities, and our encounter with nature has the additional feature of being our sole contact with worth and beauty independent of human activity. We need friends not merely as our instruments, but for what they are in themselves, and, moving one order beyond this, we need wild nature precisely because it is a realm of values that are independent of us.[42]

Following this passage, Rolston provides examples of things in nature that inspire the experiences of awe, beauty, and sublimity in us. Finally, consider his statement in *Environmental Ethics*:

> Noticing that humans value most natural things by making them over resourcefully but value a limited number of wild things as they are in themselves, we say that humans are making instrumental uses of the former type of resource but are valuing the latter type intrinsically. That is, humans may value sequoias as timber but may also value them as natural classics for their age, strength, size, beauty, resilience, majesty.[43]

These quotations bespeak pervasive errors in Rolston's arguments. In the first place, they bespeak a confusion between the *intrinsic moral value* Rolston is trying to find in nature, not just in sentient beings, and *aesthetic value*. This confusion is based in part on his reliance on the fact that humans have not created the aesthetic objects one finds in nature, and in part on his invoking the fact that experiencing nature does not use it up, or is nonconsumptive (cf. his example of viewing flowers versus making a bouquet out of them).[44] Regarding the latter point, Rolston seems to assume that whatever serves as instrumentally valuable to humans is consumed, which is surely not the case. Gravitational force (whatever it may be) is surely instrumentally valuable to human life (if it ceased to function, so would we), yet it is not consumed. Dictionaries have instrumental value to humans, yet our taking advantage of that value does not use them up.

More importantly, just because a natural (or an artificial) object produces aesthetic experiences in us, it does not follow that such an object is intrinsically valuable in the sense of possessing value within it independently of its relation to a valuer. We say humans possess intrinsic value – indeed, they are the paradigm case of things possessing intrinsic value or being ends in themselves – because they value themselves and what happens to them regardless of whether they are valued by others. On the other hand, that which is valued aesthetically logically entails a dyadic relationship with another, that is, he or she who experiences the value. Aesthetic objects are in fact paradigmatic examples of instrumental value, that is, value for producing certain experiences in others, albeit nonconsumptive instrumental value. The point is especially clear when we consider the passage quoted earlier wherein Rolston compares the intrinsic value of nature to that of religion – surely a phenomenon that depends for its value on its adherents or students.

At the risk of sounding like a Wittgensteinean, I would venture to suggest that Rolston's attribution of intrinsic value to nature in such passages rests on an equivocation on *intrinsic*. In ordinary language (or even in philosophical language), if we use the phrase *intrinsic value* at all, we use it in two separate senses:

a. To apply to that which is worth doing or having by an agent for its own sake, rather than for some other result (i.e., noninstrumental).
b. To apply to that which is not dependent on a relationship with anyone else to create value. (Only conscious beings can have value *intrinsic* to them in this sense, since only conscious beings are self-valuing.)

In sense (a), we might say that learning to play the piano is intrinsically valuable, that is, worth doing for its own sake, even if one will never get good enough to play for an audience. Implicit in this locution is in fact sense (b), by virtue of the fact that the person who is learning to play is capable of valuing that experience in himself, even if no one else does.

It is this sense (b) that we intend when we speak of humans as having intrinsic moral value – they are capable of caring about what happens to them, whether or not anyone else does. Sense (a) does not betoken any such moral dimension, it rather means that, if one is asked why one is doing something, it is appropriate to say, "for its own sake" and cut off further "why" questions. (This point is made by Plato and Aristotle.)

Thus, when Rolston says that nature has intrinsic, because aesthetic, value, what he may mean is that the aesthetic experiences one gets from nature are, like other aesthetic experiences, not a means to a further end. That, however, does not show that nature is intrinsically valuable in the moral sense, that is, worthy of the same sort of moral respect as a self-valuing human. In sum, in Argument 1, Rolston confuses aesthetic instrumental value for sentient humans, with intrinsic moral value.

Argument 2: Rolston seems somewhat aware of this concern and specifically tries to disavow an "anthropogenic" or relational theory of intrinsic value.[45] In his view, although such value is "tethered" to human experience, it is still in nature, not in us:

> A thoroughgoing value theory in environmental ethics is more radical than this; it fully values the objective roots of value with

or without their fruits in subjectivity. Sometimes to be radical is also to be simpler. The anthropogenic theory of intrinsic value strains to insist on the subjectivity of value conferral while straining to preserve the object with all its properties. It admits that the exciting object is necessary for generating value. Surely this is a paradigm beset by anomalies, ready for overthrowing. A simpler, less anthropically based, more biocentric theory holds that some values are objectively there – discovered, not generated, by the valuer. A fully objective environmental ethics can quite enjoy a "translator" when subjective appreciators of value appear. It can value such appreciation (experienced respect) more highly than untranslated objective value. Value appreciates (increases) with humans. But such an ethic does not insist upon a translator for value to be present at all, else it commits a fallacy of the misplaced location of values.[46]

But when one looks at his attempts to unpack this claim, they do not appear to provide defensible reasons to believe that the sense in which values are in nature justifies his view that nature has intrinsic moral value. Leaving aside the various confusions just detailed, one can argue that Rolston's account commits the genetic fallacy and, even ignoring that, if correct shows that virtually anything can have the value he wishes to find uniquely in nature.

The pivotal discussions of these matters can be found in Rolston's section on "Natural Value" in his *Environmental Ethics*. There, once again arguing for intrinsic value in nature, he explains, in essence, that since nature itself creates us, who are part of nature as well as valuers of nature, nature must contain within it the elements of objective value:

From a short-range, subjective perspective we can say that the value of nature lies in its generation and support of human life and is therefore only instrumental. But from a longer-range, objective perspective systemic nature is valuable intrinsically as a projective system, with humans only one sort of its projects, though perhaps the highest. The system is of value for its capacity to throw forward (pro-ject) all the storied natural history. On that scale humans come late and it seems short-sighted and arrogant for such latecomers to say that the system is only of instrumental value for humans, who alone possess intrinsic value, or who "project" intrinsic value back to nature. Both of these are inappropriate responses. The only fully responsible behavior is to seek an appreciative relationship to the parental environment, which is projecting all this display of nature.[47]

This argument seems to amount to this: We find value in nature. We are at the same time a product of nature. Thus our valuing ability is itself created by nature, both in nature's creating us and in nature's possessing the objective conditions to be valued.

In my view, this argument represents a sophisticated example of the genetic fallacy. The cause of something valuable, or capable of experiencing value, is not necessarily itself valuable, and it is certainly not necessarily morally valuable. The Nazi Holocaust may have been causally responsible for the establishment of the state of Israel. Even if we view the chartering of Israel as a morally valuable event, we are certainly not compelled or likely to view the Holocaust as similarly morally valuable.

Further, even if we ignore this fallacy and allow the argument, it simply proves too much, for it would then follow that any event that is part of the causal chain involved in producing a valuing locus such as a human, must itself possess the intrinsic moral value Rolston wishes to attribute only to nature. If the squalor, poverty, corruption, and ignorance of a city are responsible for overpopulation of humans in that city, and humans are intrinsically morally valuable, so too must be the squalor, poverty, corruption, and ignorance. If a violent rape results in the birth of a person, the rape must be valuable in Rolston's sense. In sum, Rolston's Argument 2 involves going from the fact that nature creates sentient, morally valuable creatures to the conclusion that nature is more valuable than its products.

Argument 3: Finally, Rolston seems to argue by analogy between nature and humans. Nonsentient nature is creative, self-preserving, self-repairing, dynamic, adaptable, all of which are evidenced by natural history.[48] For example, referring to organisms, he asserts that organismic value (what he calls "value$_{or}$") is a kind of intrinsic value.

Value is not just an economic, psychological, social and political word, but also a biological one. Value$_{or}$ is what is good for an organism, and all preferences and goods of humans are really subsets of this more comprehensive notion. Various instrumental organic and environmental goods contribute to an organism's well-being, and that well-being is for the organism a telic end state, an intrinsic value, not always a felt preference. Survival value lies at the core of evolutionary adaptation. Genetic information is of high organismic value, but has no necessary connection with sentience, experience, preference satisfactions, or markets. Wild creatures defend their lives as if they had goods of their own. An organism grows, repairs its wounds, resists death, and reproduces. Every genetic set is in this sense a (nonmoral) normative set, proposing what *ought to be* beyond what *is*. At this level, wild nature is a place of values prior to human decisions, and one thing the reforming world view asks is whether any concern for wild organismic value limits human decisions about land use.[19]

The problem with this discussion is patent. It shows metaphorical connections between what we feel is worthy of moral valuing in humans and nonsentient nature, and it concludes that nonsentient nature is worthy of similar moral valuing. What the argument fails to do, of course, is overcome the presumption that it is sentience that provides the morally relevant characteristic essential for being intrinsically morally valuable, and that without sentience, the other traits common to humans and nonsentient nature are irrelevant. In sum, Argument 3 moves illegitimately from the presence of some analogies between nonsentient nature and sentient creatures to the imputation of the same sort of moral status to both.

Thus, neither traditional morality nor Rolston's attempt to construct a new environmental ethic offers a viable way to raise the moral status of nonsentient natural objects and abstract objects so that they are direct objects of moral concern on a par with or even higher than sentient creatures. Ordinary morality and moral concern take as their focus the effects of actions on beings who can be helped and harmed, in ways that matter to them, either directly or by implication. If it is immoral to wreck someone's property, it is because it is someone's; if it is immoral to promote the extinction of species, it is because such extinction causes aesthetic or practical harm to humans or to animals or because a species is, in the final analysis, a group of harmable individuals.

There is nothing, of course, to stop environmental ethicists from making a recommendation for a substantial revision of common and traditional morality. But such recommendations are likely to be dismissed or whittled away by a moral version of Occam's razor: Why grant animals rights and acknowledge in animals intrinsic value? Because they are conscious and what we do to them matters to them. Why grant rocks, or trees, or species, or ecosystems rights? Because these objects have great aesthetic value, or are essential to us, or are basic for survival? But these are paradigmatic examples of instrumental value. A conceptual confusion for a noble purpose is still a conceptual confusion.

There is nothing to be gained by attempting to elevate the moral status of nonsentient natural objects to that of sentient ones. One can develop a rich environmental ethic by locating the value of nonsentient natural objects in their relation to sentient ones. One can argue for the preservation of habitats because their destruction harms humans and other animals; one can argue for preserving ecosystems on the grounds of unforeseen pernicious consequences resulting from their destruction, a claim for which much empirical evidence exists. One can argue for the preservation of animal species as the sum of a group of individuals who would be harmed by its extinction. One can argue for preserving mountains, snail darters, streams, and cockroaches on aesthetic grounds. Too many philosophers forget the moral power of aesthetic claims and tend to see aesthetic reasons as a weak basis for preserving natural objects. Yet the moral imperative not to destroy unique aesthetic objects and even nonunique ones is an onerous one that is well ingrained into common practice – witness the worldwide establishment of national parks, preserves, forests, and wildlife areas.

Rather than attempting to transcend all views of natural objects as instrumental by grafting onto nature a mystical intrinsic value that can be buttressed only by poetic rhetoric, it would be far better to nurture public appreciation of subtle instrumental value, especially aesthetic value. People can learn to appreciate the unique beauty of a desert, or of a fragile ecosystem, or even of a noxious creature like a tick, when they understand the complexity and history therein and can read the story each life form contains. I am reminded of a colleague in parasitology who is loath to destroy worms he has studied upon completing his research because he has aesthetically learned to value their complexity of structure, function, and evolutionary history and role.

It is important to note that the attribution of value to non-sentient natural objects as an instrumental and relational property arising out of their significance (recognized or not) for sentient beings does not denigrate the value of natural objects. Indeed, this attribution does not even imply that the interests or desires of individual sentient beings always trump concern for nonsentient ones. Our legal system has, for example, valuable and irreplaceable property laws that forbid owners of aesthetic objects, say a collection of Vincent van Gogh paintings, to destroy them at will, say by adding them to one's funeral pyre. To be sure, this restriction on a person's right to dispose of his own property arises out of a recognition of the value of these objects to other humans, but this is surely quite sensible. How else could one justify such a restriction? Nor, as I said earlier, need one limit the value of natural objects to their relationship to humans. Philosophically, one could, for example, sensibly (and commonsensically) argue for preservation of acreage from the golf course developer because failure to do so would mean the destruction of thousands of sentient creatures' habitats – a major infringement of their interests – while building the golf course would fulfill the rarefied and inessential interests of a few.

Thus, in my view, one could accord moral concern to natural objects in a variety of ways, depending on the sort of object being considered. Moral status also would arise from the fact that humans have an aesthetic concern in not letting a unique and irreplaceable aesthetic object (or group of objects) disappear forever from our *Umwelt* (environment). Concern for wilderness areas, mountains, deserts, and so on would arise from their survival value for sentient animals as well as from their aesthetic value for humans. (Some writers have suggested that this aesthetic value is so great as to be essential to human mental/physical health, a point perfectly compatible with my position.[50])

Nothing in what I have said as yet tells us how to weigh conflicting interests, whether between humans and other sentient creatures or between human desires and environmental protection. How does one weigh the aesthetic concern of those who oppose blasting away part of a cliff against the pragmatic concern of those who wish to build on a cliffside? But the problem of weighing is equally thorny in traditional ethics – witness lifeboat questions or questions concerning the allocation of scarce medical resources. Nor does the intrinsic-value-of-nature approach help in adjudicating such issues. How does one weigh the alleged intrinsic value of a cliffside against the interests of the (intrinsic-value-bearing) home builders?

Furthermore, the intrinsic value view can lead to results that are repugnant to common sense and ordinary moral consciousness. Thus, for example, Rolston has discussed the case of humanly introduced goats, once domestic but now feral, on San Clemente Island who were eating an endangered species of plant. According to Rolston, if one couldn't stop the goats from eating the plants, it would be not only permissible but obligatory to kill the animals in order to protect the plant, because in one case we would lose a species, in another "merely" individuals.[51]

As it happens, Rolston and I are colleagues and on one occasion had a debate on this and related issues. In the course of the discussion, I asked him to consider an imaginary case, one in which the endangered plant was threatened not by animals but by teenage trail bikers who persisted in driving over it. "Shoot them as well," he replied instantly. "There are plenty of them." This, of course, reveals quite clearly how far removed the intrinsic value position, at least in Rolston's version, is from common consensus morality, a point I consider devastating to the theory but that does not disturb Rolston. Instead he accuses me of "conservatism," whilst he is charting new moral ground.

In my view, the above case has a less paradoxical resolution. Destruction of a group of plants does not matter to the plant, whereas goats presumably care about living or being injured. Therefore, one would give prima facie priority to the goats. This might presumably be overridden if, for example, the plant were a substratum from which was extracted an ingredient necessary to stop a raging, lethal epidemic in humans or animals. But such cases – and, indeed, most cases of conflicting interests – must be decided on the actual occasion. These cases are decided by a careful examination of the facts of the situation. Thus, our suggestion of a basis for environmental ethics does not qualitatively change the situation from that of current ethical deliberation, whereas granting intrinsic value to natural objects would leave us with a "whole new ball game" – and one for which we do not know the rules.

After this long excursus, let us return again to genetic engineering of animals. Both the environmentalism that has become pervasive in society and environmental philosophy probably contribute to the mind-set that views the genetic engineering of animals (or, for that matter, of anything – plants and microbes included) as intrinsically wrong. Social environmental thought is concerned that species not disappear at human hands, holding that it is just wrong (intrinsically wrong) for us to destroy species. (I would argue, as above, that it is instrumentally wrong.) As we remarked earlier, it is psychologically a small step (but logically a vast one) to move from the claim that species should not vanish at human hands to the more dubious claim that species should not change at human hands.

As far as environmental philosophy is concerned, the specter of God, or reified and deified evolution, lurks behind some of the theorists – Rolston, for example. If nature is the product of God's (or something's) infinite wisdom, and if it is intrinsically valuable, and if it is a great systemic whole, it is easy to deduce that we, with our pitiful intellects, should not tinker with species, for we cannot alter a part without altering the whole.

Not surprisingly, Rolston quite consciously reifies species and, as I indicated in the case of the San Clemente goats, values them more than individuals. He asserts that "species *is* a bigger event than individual interests or sentience."[52] He further refers to destruction of species as "superkilling," which is morally worse than the killing of individuals – even the same number of individuals if they didn't exhaust the species.[53] Furthermore, he argues that we are morally bound not merely to preserve species but to preserve them in the natural, biological, evolutionary-ecological context in which they developed.

> A species is what it is inseparably from its environment. . . . It is not preservation of *species* but of *species in the system* that we desire. . . . The full integrity of the species must be integrated into the ecosystem. Ex situ preservation, while it may save resources and souvenirs, does not preserve the generative process intact. Again, the appropriate survival unit is the appropriate level of moral concern.[54]

Such a position, of course, strongly disposes one against genetic engineering – our wisdom cannot begin to compare to natural wisdom. So Rolston would presumably clearly resist genetic engineering of anything that could conceivably affect nature, though he does not address this question in his writings. Furthermore, even though he applauds the proliferation of species in nature as intrinsically valuable,[55] this would presumably not apply to humanly created species, since they would be artificially imposed on nature, not be "projected" by nature, and thus would not fit "in the system." It is not, however, clear that Rolston would oppose genetic engineering of domestic animals, which he describes as "captured in culture,"[56] and thus sees as radically distinct from nature and not covered by our moral obligations to "natural species."

Note in conclusion that I am not, of course, despite my

criticisms of environmental philosophy, suggesting that we should be cavalier about altering nature. I am merely suggesting that circumspection in this area – including circumspection about genetic engineering – is not a result of the intrinsic wrongness of such alteration but of the possibly untoward consequences that might result. I shall indeed devote an entire chapter to laying bare such well-founded concerns. But it is worth noting that humans have been altering nature since they crawled out of the primordial ooze; for better or worse, it is what we do, even as fish swim and birds fly.

"MIXING HUMAN AND ANIMAL TRAITS"

There is another concern sometimes voiced by those who allege the intrinsic wrongness of genetic engineering of animals. Theological types, in particular, as we saw earlier, object to the mixing of human and animal traits. Presumably, this means the insertion of human genetic material into animals, or the insertion of animal genetic material into humans. The former has occurred; as I shall discuss later, the human growth hormone gene has been inserted into animals in order to create animals of greater size. To my knowledge, the latter has not been attempted.

What, precisely, is intrinsically wrong with such an admixture? We have certainly inserted animal parts into the human body to treat disease – pig heart valves and pig skin for example – and routinely use animal products for medicinal purposes. Suppose an animal were found that contained genes for preventing cancer – sharks, allegedly, are tumor-free. Suppose, further, that this gene produced no untoward effects in humans; indeed, its only effect was to confer immunity against neoplastic disease. Why would such gene transfer be wrong in and of itself? Similarly, if the case were reversed and the gene were transferred in the other direction, from humans to animals to instill disease resistance, it is difficult to see why this would be morally problematic. To be sure, when the human growth hormone gene was transferred to animals, it caused animal disease and suffering, and thus such transfer was wrong because of its effects. But one has still not shown that in and of itself such transfer is wrong.

In a later discussion, I will raise the interesting case of transferring the genes that cause human genetic disease to animals in order to create models for human genetic diseases, which models can easily be studied and upon which experimental manipulations can be done. Many people see this as a way to eventually cure human genetic disease. Others feel that it is wrong because of the vast amounts of pain and suffering the animals will undergo if the transfer is successful. But, once again, the moral dimension of this debate seems to arise when one considers the consequences of genetic transfer, not the transfer in and of itself.

Finally, there is what one may call "the floodgate objection." Paul Ramsey has argued that genetic engineering gives us "boundless freedom"[57] by which we can create unlimited changes in life itself. Ramsey resists comparisons between previous technological changes and genetic engineering, for genetic engineering essentially allows us to play God. To the question of why this is wrong, he replies that we have not manifested much wisdom in deploying technologies developed in the past.

> Mankind has not evidenced much wisdom in the control and redirection of his environment. It would seem unreasonable to believe that by adding to his environmental follies one or another of these grand designs [for genetic engineering] . . . man would then show sudden increase in wisdom. . . . No man or collection of men is likely to have the wisdom to rule the future in any such way.[58]

Perhaps Ramsey is correct but, again, even if he is, this does not show that genetic engineering in and of itself is inherently wrong. What it does show is that humans, given our track record, will probably do more harm than good with any powerful new tool. It does not show that conceptually genetic engineering is wrong because it entails doing harm. If humans were in fact to do only good with genetic engineering, this argument would be irrelevant. In the next chapter, I will discuss strategies for minimizing harm that could possibly result from genetic engineering.

In a related vein, one often hears that if we genetically engineer animals, we will inevitably begin to genetically manipulate humans. While this is again not necessarily the case, it is certainly likely. The question of this chapter then resurfaces. What is inherently wrong with genetically engineering humans? Indeed, given the myriad foul-ups that a far from providential nature has bestowed upon the human genome in the form of genetic diseases (the standard textbook of human genetic disease lists over three thousand such diseases in three thousand pages), which often cause horrendous suffering in children, it is hard to see why we should assume that Nature always knows best. To be sure, we can always cause worse problems with our interventions, but what if we don't? Indeed, it seems to me quite likely that biotechnology will give us a purchase on many genetic diseases and their treatments – it has already done so. If we do succeed in either treating such diseases or eliminating them in the genome itself, it is very hard to see why this is wrong.

Cystic fibrosis, a genetic defect, is now being treated by use of inhalers that bring functional versions of the defective gene and its products to the lung. This way of supplying the products that are lacking due to inherent genetic defects is called gene therapy, or somatic gene therapy, in that the genetic defect inherent in the person's genome is not replaced or repaired, but rather its pernicious effects are masked. The affected person still carries the defective gene and can pass it on to his or her progeny. Germ-line therapy, which has not yet been done, would remove, repair, or replace the defective gene at the embryonic level. (Indeed, research at this level has clearly been retarded by the first version of "the Frankenstein thing," affirming the inherent wrongness of tampering with the human genome.) Yet, in fact, if these therapies succeed, not only are they not wrong, they are breakthroughs to be lauded. One can object to somatic gene therapy on eugenic grounds in that it does not cure the defect but only masks its pernicious symptoms, and thus it allows the carrier of the defective gene to pass it on whereas, in the past, he or she might have died and not reproduced. Although there may be some truth in this claim, it would certainly not apply to the therapy at the level of rectifying the defect in the gene itself. In any case, the argument is a consequentialist one and again does not show the inherent wrongness of genetic manipulation.

CONCLUSION

In sum, we have examined in detail (indeed, the reader may feel that I have provided too much detail) a welter of possible interpretations of the claim that genetic engineering of animals is inherently wrong, intrinsically wrong, something humans were not meant to do. We have found that such claims, when unpacked, fall into one of two categories. Either they involve religious appeals that cannot be translated into secular moral terms, appeals to portentous but vacuous notions like "desacralizing nature" or "breaching species barriers," and appeals to questionable metaphysical categories; or else they turn out to be claims that genetic engineering is wrong because it will have bad consequences or cause great harm. The arguments for the former category can be put behind us, but the claim that genetic engineering of animals is wrong because it can or will inevitably generate bad consequences, though very different from the claim that genetic engineering is intrinsically wrong, is highly troubling. Indeed, all cogent accounts of the first version of "the Frankenstein thing" come down to arguments about danger to humans, animals, or nature that could eventuate from genetic engineering. Thus, from the first version of "the Frankenstein thing" we must now turn to the second, the idea that genetic engineering of animals, although while not wrong by its nature, is wrong because it is likely to produce significant harm. This thesis will occupy our concern in the next chapter.

CHAPTER 2

Rampaging monsters

THE ISSUE OF POTENTIAL DANGERS ARISING
FROM GENETIC ENGINEERING

There is a thought-provoking story by Ray Bradbury about a
time in the future, in which people have mastered the ability
to travel into the past.[1] This ability is commercially exploited,
and a lively business in touring has been established. Trav-
elers are warned that they must stay on a special path provided by the company, which allows them to view events as
they occur, while avoiding any interaction with the era to
which they have traveled. In the story, a man has traveled to
the age of the dinosaurs. Despite the injunction not to leave
the path, he does so, and kills a butterfly. When he returns to
his own time, he is no longer a free man living in an advanced
society, but a political prisoner jailed under wretched conditions of totalitarian rule.

Initially, the story seems far-fetched, yet upon reflection,
one realizes that the point of the narrative is to illustrate the
way in which apparently insignificant events can create rip-
ples that, over time, can have major and unforeseen consequences. Most of us have experienced this in our own lives.
We can all recall situations like forgetting our umbrella, debat-
ing whether to go back for it, deciding to do so, just missing
the elevator, taking the stairs, pausing to tie our shoelaces,
and thereby meeting the person whom we will eventually
marry. In my own life, for example, when I applied for a
Fulbright fellowship as a college senior and chose Edinburgh

University, it was largely because, as a ten-year-old child home with the flu, I had seen an old movie about Burke and Hare set in Edinburgh, and it had seemed marvelously spooky. The year I spent there as a Fulbright scholar as a result of my whimsically based decision incalculably shaped my philosophical and personal development, from underscoring my commitment to commonsense philosophy, to convincing me that I could not continue to live in New York, as I had originally planned.

This concept has direct ramifications for the second version of "the Frankenstein thing," the notion that genetic engineering of animals is wrong because of the unknown but inevitable dangers it entails. In the original Frankenstein story, as well as in many variations thereon, the scientist does his work in order to help humanity. The dictum we examined in the previous chapter – "there are certain things that are just wrong to do" – is replaced in this aspect of the myth with the maxim "there are things that are wrong to do because they must or will inevitably lead to great harm to human beings." The archetypal image of this scenario is Dr. Frankenstein's monster on a rampage – terrorizing, hurting, and killing the innocent. The identical message emerges from the old Jewish folktale of the Golem – the being created out of clay by a saintly rabbi for highly moral motives, to protect the innocent Jews of a ghetto against wanton persecution – who eventually runs amok, killing in an indiscriminate way. In all such stories, despite the scientist's noble intentions, his activity turns out to be wrong because of his unjustifiable and hubristic failure to foresee the dangerous consequences of his actions, or even to consider in advance the possibility of such consequences and take steps and precautions to limit them.

To be sure, not thinking ahead is ubiquitous among humans – hence the clichés about people who have painted themselves into corners or who are out on a limb or up the creek without a paddle. The reason that this tendency is troublesome in science and technology, of course, is that the stakes are a great deal higher. In the first place, we endanger others, not only ourselves. In the second place, the order of magnitude of risk can be far more formidable.

Added to this is the aforementioned common sense of science's tendency to argue that scientists are not responsible for the pernicious uses to which their work is put – responsibility lies with governments, politicians, corporations, and so on. Furthermore, insofar as scientific ideology has tended to stress the value-free nature of science and to schematize the role of scientists as getting on with the job, something of a "damn the torpedoes, full speed ahead" mentality tends to dominate the scientific community. Reflective forethought, ethical and prudential, is thus both ideologically suppressed and lost in the thrill of the chase – "Come Watson – the game's afoot." And this tendency is perfectly understandable; all of us share it to some extent – any human activity, pursued with genuine passion, resists retardation, even for the most rational reasons. When we are heading for the beach on our day off, we don't stop to check the tires, however wise it may be to do so.

So the net result is that the scientific community, in its zeal to know and try, secure in the belief that science is value-free, does not fret about what it is doing and must often backtrack in the wake of disasters of one sort or another. This tendency has been well documented in a variety of areas, most forcefully, perhaps, by Paul Ehrenfeld in his *Arrogance of Humanism*.[2] Even the Asilomar conference, wherein scientists voluntarily convened to create regulations for genetically engineered organisms, was for the most part motivated more by fear that failure to act in a proactive way could usher in restrictive action by Congress that would harm the science, than by genuine concern about potential dangers that could emerge from the science.[3] A similar attitude, ten years earlier, ushered in, albeit reluctantly, principles of self-regulation for research on human subjects. Here again, the scientific community was motivated primarily by fear of restrictive legislation, rather than by moral concern for subjects, this despite a crescendo of revelations in the press about morally questionable uses of human subjects in a variety of research areas.[4]

THE SCIENTIFIC COMMUNITY'S TENDENCY TO IGNORE RISKS

So there is unquestionably a tendency among scientists to ignore or minimize dangers growing out of scientific activity. The accidental (and unforgivable) release of killer bees in California from a research colony provides a good example of lack of forethought – killer bees, as their name implies, are highly aggressive and virtually unstoppable "short of nuking the whole area," as one of my scientist friends puts it. Reassurances from the scientific community about the safety of nuclear power plants provides another example of glib dismissal of dangers growing out of technology; Three Mile Island and Chernobyl have muted such reassurances. Yet another example concerns mouth-pipetting of dangerous microorganisms and substances in the lab; despite federal guidelines against this practice, it is rife on all campuses. Indeed, any person who has served on an institutional biosafety committee can tell lurid stories of cavalier disregard of fundamental safety practices by respected investigators, or by careless workmen.

There is in fact a tendency on the part of researchers to denigrate the need for any biosafety oversight; "I've always done it this way and no one has ever been hurt" is a familiar refrain. On one occasion, before it was mandatory for research institutions to have a biosafety officer, I was discussing these issues with the provost of a major research university. I was arguing that, mandated or not, having a biosafety officer charged with enforcing compliance with sound principles of biosafety was a fundamental moral obligation of any institution undertaking research. "What are you worried about?" I was told. "Nobody's been killed here yet."

A final example, directly germane to genetic engineering, will round out my point. A few years ago, I was invited to lecture on genetic engineering at a major research university that had just received $50 million from its state legislature for developing biotechnology. The funding was largely the result of the efforts of two state legislators, both farmers, who felt that strengthening biotechnology was required for the state to compete economically. The two legislators were wise; they had written into the funding the proviso that a certain percentage of the money had to be used for examining social and ethical issues engendered by biotechnology. After my talk, I attended a party for the speakers, where the two legislators were also present. One of them was visibly angry, and I immediately assumed that my talk had provoked that rage, as my talks so often do. He assured me that this was not the case and that his ire was in fact directed against the scientist in charge of the program. "Why?" I asked. "Didn't you hear his speech?" he replied. "His very first project involves working with a chemical company to test their new herbicide-resistant wheat seeds. The theory is that one can soak the earth with their herbicide, yet the seeds will grow." "So what?" I asked naively. "So what! We happen to have a ground-water contamination problem here already as a result of herbicide overuse. This will certainly add to the problem!" I began to understand his concern. "Have you approached the scientist?" I asked. "How on earth can he justify such research?" "That's why I'm so angry," said the legislator. "He said there is no cause for concern, that they would fix the problem by genetically engineering a microbe to eat the contaminants." Suddenly full understanding dawned on me. "I see," I said, "You're worried that it will also eat the state capital, and if you express this, he'll say 'Don't worry, if that happens we'll build a microorganism to puke the state capital.'"

In sum then, scientists tend to be cavalier about the dangers emerging from science and technology, as Dr. Frankenstein was, or to ignore them, and this lends credence to the second version of "the Frankenstein thing," for scientists are unlikely to anticipate potential dangers on their own.

UNANTICIPATED DANGERS IN
GENETIC ENGINEERING

This problem is amplified by a far more serious problem. Even if scientists were to zealously attempt to anticipate any and all potential dangers, and to safeguard against them, this is essentially impossible to do, by the very nature of things. We simply lack the ability to predict everything that can possibly go wrong.

This latter concern has been most eloquently and elegantly presented by the physician-novelist Michael Crichton. Many of Crichton's novels are based on extrapolations from current science and are informed by a unique combination of scientific knowledge and understanding, literary virtuosity, and a masterful ability to teach difficult concepts to lay persons. In his novel *Jurassic Park*, Crichton engages the issue of unanticipated (and unanticipatable) dangers growing out of genetic engineering of animals.[5]

In the course of the novel, Crichton orchestrates some of the main themes we have been discussing, including scientists' distancing themselves from ethics through their ideology and the "damn the torpedoes" attitude that leads investigators to try things without deeply concerning themselves about potential consequences. Crichton's basic point, however, is that even when scientists do attempt to provide failsafe mechanisms for potential dangers arising out of genetic engineering of animals (or, indeed, any radically new technology), these safeguards are extremely unlikely to be adequate. The reason for this is that such safeguards are based on those problems that can be anticipated, yet the really problematic cases must inevitably involve the sorts of situations that *cannot* be anticipated.

Crichton's fictional vehicle for making his points is the prospect of re-creating dinosaurs by genetic engineering for a dinosaur wildlife park. Although scientists build into both the animals and the park various clever mechanisms to prevent the animals from reproducing or leaving the confines of the preserve, things do not go as planned. The conceptual point underlying Crichton's story is drawn from the relatively new branch of mathematics known as chaos theory, which postulates that the sorts of intrinsically predictable systems beloved by Newtonians, determinists, and introductory philosophy professors – the billiard table whereupon, if all forces applied to the balls are known and the initial conditions are described, one can predict with certainty the movement of the balls – are few and far between, and, in any case, are essentially irrelevant to complex new technologies like genetic engineering. Crichton puts his point in the mouth of a mathematician named Malcolm.

> "Jesus," Gennaro said. "All I want to know is why you think Hammond's island [the dinosaur preserve] can't work."
>
> "I understand," Malcolm said. . . . Simple systems can produce complex behavior. For example, pool balls. You hit a pool ball, and it starts to carom off the sides of the table. In theory, that's a fairly simple system, almost a Newtonian system. Since you can know the force imparted to the ball, and the mass of the ball, and you can calculate the angles at which it will strike the walls, you can predict the behavior of the ball far into the future, as it keeps bouncing from side to side. You could predict where it will end up three hours from now, in theory."
>
> . . .
>
> "But in fact," Malcolm said, "it turns out you can't predict more than a few seconds into the future. Because almost immediately very small effects – imperfections in the surface of the ball, tiny indentations in the wood of the table – start to make a difference. And it doesn't take long before they overpower your careful calculations. So it turns out that this simple system of a pool ball on a table has unpredictable behavior."
>
> . . .
>
> "And Hammond's project, . . . is another apparently simple system – animals within a zoo environment – that will eventually show unpredictable behavior. . . . The details don't matter. Theory tells me that the island will quickly proceed to behave in unpredictable fashion. . . . There is a problem with that island. It is an accident waiting to happen."[6]

Chaos theory in fact represents a formalization of a number of points well-known to common sense but that often escape the common sense of science. First of all, as any mechanic will readily attest, "things don't work like they're supposed to." There is a major gap between theory and practice. Call it gremlins, call it Murphy's Law – "shit happens," as a popular bumper sticker reminds us. Second, insofar as any system must be implemented by humans, a weak and unpredictable link is ipso facto present. Third, when scientists talk about Newtonian predictability, they are talking about closed systems where everything is tightly controlled and all relevant variables known and variations in initial conditions – imperfections in the pool table, in the above example – are abstracted from. In the real world, this does not occur. Not only are there imperfections in the table, windows fly open and gusts of wind blow the balls askew, cockroaches crawl up on the table and block the movement of the ball, demented snipers shoot the ball, cats attack the ball, ceilings collapse, and so forth.

It is precisely this sort of insight that leads common sense to take what experts say with a grain of salt. "I don't care what the experts say," is often the response of the ordinary person living in an atmosphere that smells funny; "breathing that stuff can't possibly do you any good."[7]

Part of the conflict between science and common sense in this sort of arena has to do with differences in canons of evidence and justification – the scientist will never say that the funny smell is harmful until he or she has done exhaustive epidemiological work to confirm that there is a problem and has confirmed the operative mechanism that causes the problem. Common sense – most certainly a mechanism for survival – demands far less and is content with affirming on the basis of vague experience of the race that "what don't smell right ain't right." Scientists make careful, complex judgments, and they attempt to do this in a detached way; common sense's judgments have no pretense of disinterestedness. Here Hume is correct; one must work to be a philosopher or scientist, but one's humanity comes naturally.[8] Thus ordinary people remain fearful of buildings in which disease research is performed – especially on nightmarish pathogens like the AIDS virus, the tubercle bacillus (TB), the rabies virus – despite scientific assurance that adherence to federally mandated guidelines for containment assures that there is no danger.

At any rate, ordinary common sense is very much in accord with Crichton's graphic portrayal of scientifically unanticipated dangers growing out of genetic engineering. And the common sense of science's insistence that science plays by its own rules – for example, demanding agnosticism about animal consciousness because we can't experience it – has led to ordinary people's increased skepticism about the pronouncements of experts, who are often perceived as out of touch with reality and marching to their own drummer.

The sort of response that Crichton's book engendered in a leading trade journal, *Genetic Engineering News*, helps to explain why "trust me, I'm a scientist" does not work very well to reassure the public. In an editorial accompanying an article about Crichton's book, the publisher warns that though members of the biotechnology industry might scoff and feel that no one would take such a story seriously, "How about the group of community activists that successfully halted the field tests of frost-resistant crops in California? Or another group that successfully thwarted the construction of a biotechnology facility in Massachusetts?"[9] The editorial goes on to argue that material like Crichton's requires that the biotechnology community "educate the general public" about biotechnology, about "what it can do, and what it cannot do."[10]

The article itself is entitled "Will 'Jurassic Park,' the Movie, Create a PR Problem for Biotechnology?"[11] In the article, the author quite fairly summarizes numerous concerns expressed by Crichton in his book: Many biotechnology workers are motivated by profit and are thus not disinterested scientists; scientists tend to avoid thinking about the ethical implications of their work and to deny responsibility for untoward outcomes the work engenders; the biotechnology community has proceeded in great haste in a thoughtless fashion; adequate regulatory oversight of biotechnology does not exist; and, most important, potential dangers are not adequately dealt with.

The main concern of the article is expressed in its title. To answer the question, the author polled a sample of leaders in the biotechnology industry, most of whom indicated that they had encountered no public concerns growing out of the book. One person indicated that he had received requests about whether Crichton's fictional scenario was possible. His reply, as quoted in the article, is that "Crichton has taken current knowledge and moved via theoretical extension to future possibilities. In doing so, he has not performed sound science, but he has displayed good fiction."[12] The article's author indicates that most industry representatives believe that the public can distinguish fact from fiction. Others stressed the impossibility of in fact making dinosaurs.

The reader should marvel at the article in the context of our discussion as it, in essence, tends to miss many of Crichton's points and to instantiate others. In the first place, Crichton clearly did not write the book to alarm people about dinosaurs, or to crusade against the particular activity depicted in the novel. He is rather using this fictional example as a vehicle to illustrate some genuine problems endemic to genetic engineering – lack of regulation, fools rushing in, pecuniary motivation of many biotechnologists and companies, and, most important, the inability to anticipate all possible dangers of any frontier technology, and the consequent need to be exaggeratedly circumspect. In short, he is presenting, in palatable form, an occasion for asking questions that must be asked – the sort of questions we are asking in this book. He is not antiscience and not antibiotechnology; indeed, in a box accompanying the article, he is quoted as follows: "Biotechnology is absolutely astounding, but I wanted to sound a cautionary note. Science is wonderful, but it also has its hazards. If the book makes some people uneasy, maybe it should."[13] Furthermore, albeit incidentally, why should Crichton be attacked for what is essentially a plausible extrapolation from current science? Such extrapolation, in a positive vein, is practiced all the time by the industry to promote public acceptance of biotechnology – witness the promise to cure human genetic disease by biotechnology; this is certainly plausible, but still technically distant.

Most important though, neither the editorial nor the article engage any of the real issues raised by Crichton, nor do they demand that the issues be raised in the biotechnology community. The genuine issue of controlling potential danger, as well as the other issues I have mentioned, are not treated as genuine issues; the only concern discussed is the potential PR problem!

And the solution suggested by the pieces? To "educate" the public. What does that mean? It does not appear to mean to really engage the issues in a public forum; rather it seems to mean to assure the public that dinosaurs won't be made and thus won't run amok. If the intention was to genuinely educate, the magazine (or other parts of the industry) would surely begin by running frequent, intensive articles on the welter of genuine and spurious ethical and prudential issues associated with biotechnology, or to demand forums for doing so within the biotechnology community itself. Instead, the magazine indicates only its intention to establish a speaker bureau so that the public can have access to the most authoritative and respected members of the field. In short, education seems to mean telling and reassuring, not probing and dialogue.

I do not intend to pick on this magazine or this article; I in fact read *Genetic Engineering News* regularly and with pleasure, and I learn a great deal from it. Rather, I use this example to buttress my earlier claim that the biotechnology community, as a microcosmic reflection of the scientific community in general, does not typically involve itself in ethical issues in its field except insofar as they become public relations issues and, even then, does not truly engage the issues. (There are notable exceptions to this generalization, which I will chronicle as I proceed.) Crichton is in fact doing what the field should have been doing long before he did!

NEW TECHNOLOGY AND RISK ASSESSMENT

That there are potential dangers – perhaps significant dangers – associated with genetic engineering of animals is unquestionable. The biotechnology community seems to believe that such an admission will jeopardize its existence by fueling public fear. In actual fact, exploring the possible dangers, and delineating the mechanisms by which they can be checked, are not only morally obligatory but prudentially wise as well. Radio is more frightening than television; people's imaginations can generate scenarios even more frightening than real ones. In the event that something goes awry, people are likely to be outraged more if they have been fed patronizing pablum assurances that biotechnology presents no danger, than if they have been told the truth and kept informed of the safeguards implemented to minimize the dangers.

No new technology is risk-free – even if a tool has overwhelmingly positive advantages, it can and will usher in disadvantages as well, some predictable, some not predictable. Accidents were a predictable risk associated with the automobile when it was introduced; air pollution and dependence on foreign oil were not. Similarly, no one appears to have predicted that in addition to saving lives by killing pathogenic bacteria antibiotics would in essence select for bacterial strains resistant to the drug in question. And, while visionaries like Norbert Wiener saw computers and other cybernetic mechanisms as rescuing people from mundane, repetitive, tedious jobs,[14] no one seems to have anticipated the erosion of privacy that they have occasioned, by making one's financial and other data instantly accessible to those willing to pay a fee for obtaining it. In the same vein, it is difficult to blame the manufacturers of hair spray for not anticipating the effect of their propellants on the ozone. On the other hand, we can blame the makers of lawn darts for not worrying about injuries to children and the cavalier deployers of Agent Orange for their failure to anticipate (or care about) health risks to innocents.

Life itself is fraught with risk, some of which can be anticipated and managed, and some of which cannot. Astronauts visit the moon safely, yet slip in the bathtub and incur serious injury. Indigestion is a foreseeable risk in visiting fast-food restaurants; being machine-gunned by a psychotic is not. We do what we can to minimize predictable risks and, unless we are very neurotic, do not fret about the unforeseeable and freakish ones. Rational people, possessed of different values, will disagree as to what risks should be managed and at what cost. I ride motorcycles and derive great pleasure from doing so. Obviously, I risk greater chance of injury on the cycle than I would if I traveled by auto or tank. Yet I am prepared to take that risk because of the pleasure I derive from motorcycles. By the same token, consider one's eating behavior. Evidence indicates that, all other things being equal, we are likely to live a bit longer by staying thin and avoiding fats. If such a regimen increases my statistical life span by a month, is it irrational to flout it? Or is a lifetime of éclairs and french fries a fair trade for a life shorter by a few weeks? Clearly, one's value system will determine the choice.

On a social level, such individual freedom does not help a great deal in deciding policy. When nuclear plants or high-tension electrical transmission lines are placed in my area, I do not choose the attendant risks that their presence may occasion. Indeed, I may enjoy no compensatory or mitigating benefits that help balance the risks; the electric power, for example, may benefit another community.

Thus people may, indeed, often be subjected to risks they do not choose and that do not carry benefits for them. Such is the nature of social life. In a representative democracy, representatives are presumed to speak for their constituencies and to protect their interests against being subjected to severe, unmitigated risks. Thus, if a federal agency wishes to dump nuclear waste in a remote portion of a western state, legislators, governors, and their staffs can be presumed to represent the interests of their constituencies, both through official channels of recourse and through unofficial but significantly powerful forces such as the press and public opinion.

In the end, a given set of risks cannot possibly be shared equally across the population. It is more plausible to place a nuclear waste dump in a sparsely populated area than in an urban metropolitan area, thereby subjecting fewer people to the risks of accident. So the rural population, in this case, bears a significant degree of the risk associated with the nuclear industry. On the other hand, the urban population may bear a disproportionate amount of risk in a different domain – for example, that population stands at greater risk of attack by an enemy seeking first-strike advantage or by terrorists. So, in a fair society, the risks are incurred by different segments of the population in such a way as to assure that different burdens of risk are appropriately shared. (One of the worst features of colonialism, of course, is that it did not even attempt to fairly distribute risks and benefits.)

VALUES AND RISK ASSESSMENT

Contrary to what many experts would have people believe, risk analysis and minimization is far from an algorithmic exact science. As Nicholas Rescher has argued at length,[15] a host of value judgments must be invoked in assessing risk. For even if science provides a precise, quantitative, objective account of the likelihood of each possible risk, for example by studying the ratio of catastrophes, say airplane crashes, to the total number of flights, the question of judging how much to count that risk, as against what is lost by not taking the risk, is going to be a matter of values, as we saw was the case in individual decisions. Our major problem in social decision making is finding a suitable way of summing up often incommensurable social values. How, for example, do we weigh the desires of those members of the population who are willing to risk significant harm – job displacement, health risks, accident risks – for the sake of economic growth, versus those who are unwilling to do so, and who argue we should instead curb our appetites and lifestyles? How do we weigh risks to future generations who do not yet exist as against present benefits?

How do we weigh the dependence on fossil fuels, the correlative dependence on foreign countries, and the environmental despoliation associated with the use and transport of oil against the dangers potentially associated with nuclear power? How do we weigh the benefits that accrue to billions of animals by constraining factory farming against the increased cost of food products that such reform is likely to entail? All such decisions must be made by appeal to social values, including to a great extent, ethical values.

For example, to use a case I have already mentioned and will develop further in the next chapter, over the past two decades the American public has been developing new and significant moral concern for the treatment of animals in society, which concern focused first on the use of animals in biomedical research. From 1980 on, there was an increasing demand for regulation of animal use in science. The research community responded to this demand by emphasizing the risk to human health associated with any legislated constraints on the use of animals in the research process; only by laissez-faire (i.e., lack of regulation) in animal use, it was claimed, could the conquest of disease be expeditiously achieved. In assessing the evidence, the public seemed to disbelieve the latter claim and weighed the moral concern for proper animal treatment as being more exigent than the alleged danger to human health attendant to regulation, and Congress thus passed two laws in 1985 aimed at assuring research animal well-being. Thus the growing concern for animals seemed in the social mind to supersede both the traditional autonomy of researchers and their argument of risk to humans accompanying curtailment of that autonomy. As it happens, it now seems clear that the public was correct; the new legislation seems to be assuring both good science (by minimizing animal stress and pain, which can skew research results) and good animal care and treatment, which, despite its scientific impact, was always a low priority even in major research institutions. Thus a win/win situation has been created, and the public posture seems vindicated.

All policy making and all social decision making is driven by value commitments, as well as by factually judging risks and benefits. Obviously, the more certain we are of the nature of potential risks, the more easily we can evaluate them and build safeguards against them. When it comes to a radically new technology, of which genetic engineering of animals provides a paradigm case, the process associated with social decision making regarding risks and benefits is rendered significantly more difficult, as no one can begin to assess the probability of risks or benefits, let alone assign valuational weight to them, for no one is even clear as to what they are. Thus, as the OTA official told me, proponents of genetic engineering stress the possible benefits and ignore the possible risks, while opponents stress the possible risks and ignore the possible benefits. Since we have virtually no history of producing, releasing, and consuming genetically engineered animals; since we have no full understanding of how individual genes and changes therein affect the whole organism in any given case, especially at the phenotypic level; since our practice exceeds our theoretical understanding and even our intuitions, we cannot draw a confident empirical or probabilistic risk-benefit profile to which we can then apply our values. For that matter, as I have indicated throughout this discussion, we are not always clear about our relevant rational values in the case of genetically engineered animals. What, then, does one do when components vital to rational decision making are absent?

"BOOTSTRAPPING"

Let us return for a moment to individual, rather than social, decision making. Do we, as individuals, encounter situations that are analogous in their uncertainty to the social uncertainty regarding genetic engineering of animals? Surely we do – in both our personal and professional lives. Many of us, at some point in our working lives, are faced with the choice of changing jobs and locales. The locale may well be totally alien to us, it may be rural after we have spent out entire lives in a megalopolis, or vice versa. It may even be Saudi Arabia after we have spent forty years in Topeka. To further complicate matters, the job may be something for which we have not been trained and in which we have no experience. (An academic may be asked to become chief administrator for a charity, for example, or a politician may be offered a university chair.) How do we proceed as rational agents?

Sometimes, of course, we just proceed as when we suffer a midlife crisis. Arguably, when possessed by that demon, we are not in fact rational agents. But not all or even most people faced with such choices are possessed. So what sort of reasoning do we go through? Obviously, in the case of changing jobs, we research the position and the locale – by reading books, talking to people, visiting the place, and so on, to try to get a sense of what we are likely to encounter, or, failing that, what we might encounter. Without such an informational grounding, we literally cannot deploy our value system in weighing situations and emerge with a decision. To be sure, the information may be seriously limited – "Quimby's brother lived in Saudi Arabia and hated it. Or was it Egypt?" – but without some information, our machinery for rational decision making cannot even be energized. We consider the money we will make in the new job versus the money we make now, factor in the cost of living, add a credit for adventure and a debit for displacement, worry about the possible effects on the children of leaving their friends, talk it over with friends and strangers who will listen – and muddle through. Most significantly, perhaps, we constantly examine the question "What do I lose if it doesn't work out?" and construct worst-case scenarios, and we try to minimize that potential for loss. For example, we might try to get a leave from our current position and try the new one for a year, in effect looking before leaping or, better, tying a safety line to one's waist before leaping. And what do we ultimately gain by trying it for a year – after all, we have only postponed the decision. The answer is obvious – we have gained the knowledge that allows us to make an informed decision.

Can this model help us with social decisions to be made on the risks (and benefits) associated with new technologies like genetic engineering of animals? I believe it can, for a great deal of our rational, social decision making seems to involve muddling through – one book calls it "bootstrapping."[16] It is essentially what we have done with all new technologies from the wheel to the automobile and the computer; what else could we have done, given the uncertainty that, paradoxically in a way, only diminishes as one proceeds?

SKEPTICISM ABOUT "EXPERTS" JUDGING RISK

What, then, should be the components of a rational and just approach to dealing with potential risks associated with genetic engineering of animals? We must first of all realize that common sense, which must be satisfied for genetic engineering to be able to proceed as a viable technology in society, may rightly distrust "experts" when issues of risk are involved. There are two senses in which such suspicion of experts is reasonably manifested in ordinary life:

> Sense 1: Significant numbers of people in society may not trust experts a) to fully identify risks associated with their area of expertise, or even if they identify such risks, b) to take them seriously enough or weigh them heavily enough to suit common sense.
>
> Sense 2: A fortiori, many people in society may not trust experts to decide for society in general what risks members of society should take in pursuit of a particular technology or innovation, in our case genetic engineering.

A viable public policy regarding the management of risk associated with genetic engineering of animals must take cognizance of both of these reasonable components of public skepticism about leaving control of risk to experts.

As far as Sense 1, identifying risks and their likelihood, is concerned, I would argue that *brainstorming* about risks is an absolutely essential first step toward managing this technology. By brainstorming I mean throwing every conceivable risk out on the table before one begins to discuss their likelihood. "Experts" with a cognitive, emotional, and perhaps financial stake in genetic engineering are too likely to precensor what they are even willing to count as possible risks. To take a simple case that I alluded to earlier, we know that, as common sense asserts, familiarity breeds contempt. Risks that we have ceased to even think about do not leap to mind – routine mouth-pipetting of dangerous material by scientists in their laboratories is unlikely to be listed by the scientists themselves as a biohazard associated with their work, though a lay person may shudder at the very idea. Similarly, crossing the street in the middle of the block, in heavy traffic, against the light, and dodging by centimeters hurtling cabs driven by what to all appearances are frenzied maniacs would not be listed as a risk by most New Yorkers, though I have witnessed catatonically frozen tourists from the heartland unable to bring themselves to step off the curb.

In the case of scientists, other factors besides familiarity suppress or filter out cognizance of what, to ordinary people, are potentially weighty dangers. One such factor, already mentioned, is scientific ideology, which focuses scientists' attention toward unearthing truths and advancing the corpus of knowledge, with dangers to themselves or those attendant to their work relegated to the periphery of concern. From personal experience, I would argue that the sheer joy of scientific inquiry and the excitement of the chase, as much as anything else, militates in favor of getting on with it, in the same way that a child (or some adults) with a new toy does not want to stop and read the directions. For most scientists of my acquaintance, this fervor is a far greater factor in leading to the lack of reflection on risks than is the profit motive.

Nonetheless, as biotechnology research becomes ever more closely identified with private industry, private or corporate rather than public funding, and an "eyes on the financial, rather than Nobel prize" gestalt, the financial dimension becomes increasingly significant in the minds of genetic engineers. Great financial rewards await the scientist who can generate a commercially viable product, and the negative effect of money on caution is so well-known as to be the stuff of myths, legends, and Yosemite Sam cartoons. Thus, both the excitement of discovery and its potential reward can lead not so much to a "damn the torpedoes, full speed ahead" attitude as to a "what torpedoes?" blindness.

Finally, potential – to them, fanciful – risks do not mesh well with scientists' positivistic tendency to deal with what can be measured, analyzed, observed, and quantified here and now.

To many scientists, worrying about science fictionish, far-out risk scenarios is as temperamentally and ideologically irrelevant as worrying about who will be one's next-door neighbor in the afterlife. "Facts" are real; possibilities (except for empirically based probabilities) are the stuff of fiction, not reality, best left to Michael Crichton and other spinners of tales.

The factors just mentioned, plus a poor track record of failed reassurances regarding past innovations, lead society to a reluctance to trust scientists alone to identify the risks growing out of new technology. If scientists fail to see or take seriously risks that seem plain to us, their credibility as risk managers diminishes. Furthermore, many of us have witnessed numerous cases in which scientists' reassurances about the safety of new innovations and the minimal dangers to society of such innovations have turned to ashes. Experts told us that living near a nuclear plant is safer than taking a bath – this was dramatically belied by Three Mile Island and, more tragically, by Chernobyl. Experts told us that "there is no scientific evidence" that living near electrical transmission lines poses a danger to health, yet reports of the high incidence of leukemia in children who live near such lines continue to roll in.

We can all recall numerous examples of failure of expert reassurance or cases where experts have blundered or been blindered. Who among us can forget the space shuttle tragedy, when all the alleged fail-safe mechanisms failed, and the shuttle was launched on a freezing cold day? Who among us has not known a tragic misdiagnosis – women's shoulder pain dismissed as menopause and turning out to be lung cancer or heart disease; men's back pain dismissed as a charley horse and turning out to signal prostate cancer – despite our formidable armamentarium of diagnostic tools?

In the early 1980s, I personally experienced an incident of expert failure – failure to acknowledge the reality of a risk and the correlative dissemination of reassurance to a community that it was not at risk. When our university was contemplating the undertaking of AIDS research, members of the community were fearful of the possible risks to the general population that might emerge from such research. As it happens, there is a branch of the Centers for Disease Control (CDC) located at our university, and, when a major AIDS expert from another CDC research center came to the community, he agreed to speak to the public about the risks stemming from researching the virus. Such risks, he assured a large audience, were minimal – virtually nonexistent. One needed "to work" he said, to catch the virus. Outside of engaging in venereal contact with an AIDS sufferer or receiving a transfusion of infected blood, one simply could not contract the disease, he averred. There was, he said, "no scientific evidence that you can acquire AIDS in any other way." There was "no risk" of acquiring the disease in a laboratory context. "Why," he declared, "I would *bathe* in the AIDS virus – there is no danger of catching it that way." Within six months of his skeptically received pep talk, the first case of a person contracting AIDS through a cut in the skin was reported. Obviously, the public reluctance to trust this expert was well founded.

I learned more about public attitudes toward scientific risks during the period when our university was deciding whether or not to undertake AIDS research. I had been asked by our campus biosafety committee, of which I was a member, to play devil's advocate and make the strongest case against our undertaking such research. I agreed to do so, and, in my remarks, cited among other things, the fact that we did not have a functional incinerator for disposing of biohazardous wastes.

Much to the committee's – and my own – surprise, an enterprising student reporter from our college newspaper requested and received the minutes of that meeting. I was understandably shocked to see my remarks on the front pages of both our college and city newspapers. "Rollin says university is not a mature research institution regarding biosafety control" the headlines screamed. Needless to say, I would not have won any popularity contests among our researchers that week and, indeed, was called upon to explain my remarks before the university research council. Interestingly enough, in the course of that heated meeting, only one researcher defended my remarks – the very scientist who wanted to undertake the AIDS research! His rationale was that it is better to examine every possible problem forthrightly and openly in advance, than to learn by tragic hindsight. This argument was not well received by the other researchers, and I was beginning to entertain Gauguin-like escape fantasies. However, all of this changed on the Friday of that week. The city newspaper ran an editorial engaging the question of whether the university should embark upon AIDS research. Amazingly, their answer was a resounding yes, "as long as there are people like Bernie Rollin in the university structure who are able to articulate the interest and concerns of the community." In other words, people felt that there is a commonsense voice that needed to be heard as well as a vested-interest expert voice.

Whereas the "trust me, I'm a scientist" stance may have been reassuring to the general public in the heyday of social adulation of the sciences, it will not wash in an era of mistrust for a technology that elicits reflexive Frankenstein imagery. And it is in fact patent that the public does not fully trust expert pronouncements on the safety of biotechnology and does not even begin to understand biotechnology. For most scientists this public lack of understanding militates in favor of letting them decide the risks for the public – "how can the public judge what it does not even understand?" is a commonly heard complaint. But the fact is that this is not the conclusion that the public draws regarding biotechnology – according to a brand-new study, some 80 percent of the public wants to be consulted in a far more significant way on setting biotechnology policy.[17] Though it is not clear what, precisely, this means, it follows from our discussion thus far of the sources of expert blindness to risks that there should be extensive dialogue between scientists and the public on what risks grow out of genetic engineering, and how much these should be weighed.

Unfortunately, scientists resist what they see as the blind leading the sighted, and the government tends to concur; bureaucrats motivated overwhelmingly by the First Commandment – "Cover Thy Ass" – are drawn to experts, not to the unwashed masses. Yet it is both morally and pragmatically essential that the public be a pivotal part of the decision process, for they will incur the risks, and they will be asked to accept the results of biotechnological innovation.

I have thus far focused my discussion on Sense 1 of rational public distrust of leaving risk management to experts, specifically on the factors that can and do prevent experts from seeing risks that ordinary people might see or weighing them as heavily as ordinary people might. If public suspicion of genetic engineering is to be overcome, there must be an extensive dialectical interchange between the public and the experts on what is to count as a risk, and how that risk should be weighed. Even if the experts are always right, and public suspicion is misguided, despite the reasons I have given above to doubt that this is the case, the only rational way to change public opinion is through education, and this is best accomplished dialectically. All public concerns and suspicions should be exposed to the clear light of day, dissected, and consensus reached about how seriously they should be taken.

It could be argued – and, indeed, has often been argued – that the sort of dialogue just alluded to is impossible. The public, it is asserted, is scientifically illiterate, hopelessly gullible and naive, and cannot even understand scientifically based arguments against the reality of a certain risk. There is certainly ample evidence in support of such a claim. Vast numbers of people believe the cover stories on supermarket tabloids – that Hitler was a Martian, or that perpetual motion machines or cars that are fueled by water exist but are kept hidden away by the CIA and auto manufacturers respectively. Polls have shown that large numbers of precollege science teachers believe in the supernatural and that alarming numbers of the public doubt the reality of the Holocaust.

All of this is true – and demoralizing. Nonetheless, if one believes in democracy, one must believe in the educability of the public and in its wisdom. We do, after all, trust the public to judge matters of life, death, and property in the jury system and in voting. It is fashionable among intellectuals to deride the wisdom of the hoi polloi; yet these same intellectuals forget, as has often been remarked, that outside of our small areas of expertise, we are all the general public.

My own work for the past seventeen years has been closely involved with explaining the complex issues in bioethics to very diverse groups. I have lectured almost seven hundred times on such complex bioethical issues as animal use in science, agriculture, and genetic engineering; medical ethics; death and dying; scientific ideology; and animal pain and consciousness to such diverse groups as Rotary clubs, rodeo cowboys, church groups, slaughterhouse workers, corporate executives, bankers, ranchers, farmers, circus and fair promoters, law enforcement personnel, medical school faculty, veterinarians, attorneys, psychologists, and 4–H children. In the overwhelming majority of cases, I have found people to be open, intelligent, fair-minded, and interested, with our sessions going on for many hours past the officially designated schedule. I have furthermore sat in on many animal care and use committees and watched lay people, whose presence is required by federal law, interact with scientists, again with the same result. In this context, people must make decisions. I think most scientists on such committees would report that the lay people are generally open, educable, wise, and fair. All of this buttresses my belief that there is very little in science (excepting theoretical physics) – certainly there is very little in biological science – that cannot be explained to ordinary, interested members of the public of normal intelligence by patient, committed experts who truly enjoy teaching. Since matters of potential risk are of interest to and affect everyone, this area should especially lend itself to dialogue.

What of the second level of rational suspicion of experts (Sense 2) – whether experts should decide for all of us, or make policy for all of us, regarding the dangers growing out of their area of expertise? Assuming we have identified the areas of risk in genetic engineering and the likelihood of those risks, should not experts make the policy regarding risks we the public should take? This is, in fact, what often occurs in society – we ask researchers to determine research policy; medical people to determine medical policy.

Such deference to experts is, in fact, one of the most powerful and subtle dangers to democracy. First of all, it is overwhelmingly seductive, since the world is so complex. Let the experts in a given area, who know the facts of that area better than the rest of us ever could, make the value decisions on what risks we should take in the area and how we should regulate it. Seductive as this is, it is fallacious, since it conflates fact and value, "is" and "ought," knowledge of what is the case with how we ought to regulate the area in question. Though experts may know the facts of an area better than we do, they may have very different valuational attitudes toward these facts than the rest of us do. And, since we have seen that risk management is a matter of both facts and values, experts may, because of their value system, be inclined to weigh risks very differently than the rest of us do and be differently concerned about managing them.

Consider some examples of how this can occur. We might be prima facie inclined to see physicians as being in the best position to decide the right and wrong of medical social policy because they know most about medicine. The problem, of course, is that physicians often have values that do not reflect those of society in general. For example, doctors are much more committed to keeping people alive at all costs than are many patients – perhaps most patients. Doctors tend to emphasize cure, not care of suffering; doctors tend to see pain as a side effect of disease, not as a focal point of concern; doctors may be less concerned about high health care costs, from which they benefit, than are patients, and so forth.[18] Thus, what doctors may see as acceptable risks associated with health care policy may not be so viewed by the rest of us.

As another example, consider funding for research that is paid for by public money. At the moment, research priorities – what should and should not be researched – are dictated by panels of experts in the fields in question. One can argue, as I have in fact done, that while we need experts to help us under-

stand what the point and purpose of a given research proposal is, we, the taxpayers, should decide where we want the
money spent – it is, after all, our money.[19] We can surely seek
expert input, but the final decision should be as democratic as
possible, with as many positions about where the money is
best spent submitted as possible. Indeed, putting that decision in the hands of the experts may well have negative consequences; it may block (and in fact has blocked) radically new
approaches; it may end up with experts serving themselves
rather than the general public, and so forth. The values of
experts may not be the values of the larger society in which
they reside. (Of all the hundreds of speeches I have given to
scientific groups, incidentally, nothing evokes as much anger
as the above suggestion.)

I would argue that genetic engineering fits much the same
model. Genetic engineers are not the proper people to alone
dictate policy about genetic engineering; the public is, though
of course it cannot do so without expert factual input from the
scientists in the field. With regard to possible risks associated
with the genetic engineering of animals, the point obtains in a
number of ways.

DEMOCRATIC RISK ASSESSMENT – INVOLVING THE PUBLIC

To summarize our discussion thus far: There are, conceptually, two components to judging risks in genetic engineering. The first component involves attempting to get some
measure of the empirical likelihood of a given risk. In well-
established areas, risk probability can be determined by relative frequency; if, to the best of our knowledge, nothing has
changed, we can extrapolate the past to the future. But in a
new technology, there is no history and limited knowledge.
So we cannot even presume to be sure of what the possible
risks are, let alone attempt to assess their likelihood. We have
already seen that scientists are likely to ignore or belittle, or
reject out of hand, or set aside certain risk possibilities or

scenarios by virtue of their ideology, situation, and mind-set. Furthermore, scientists are loath to speculate.

Citizens, on the other hand, have lively imaginations and a vested interest to speculate, not ignore, when it comes to possible dangers to themselves. They are not convinced that there will be no Frankensteins, Andromeda strains, or cockroaches that will eat Cincinnati. And we know both from well-known surveys and from daily life that people do not trust scientists' reassurances as they once did. Further, if the risks materialize, the very citizens who do not see any benefit from genetic engineering could be hurt as much as or more than the scientists who are enthusiastic about it and may profit from it. (Some of the public, after all, lives in Cincinnati.)

The conclusion, then, is obvious. Governments should not, as they have traditionally done, look to scientists alone to guess at (and minimize) the possible risks associated with a new technology like genetic engineering of animals. On the contrary, they should seek, from the public, as many concerns about risk as possible, however outlandish these may seem to scientists. Since risks are unknown anyway, let us, for safety's sake, cast our net as widely and diversely as possible. Scientists can assist the public by explaining the new technology in terms accessible to the lay person. And after all fears are outlined, scientists should be encouraged to bring forth their expertise if they believe a particular concern is remote or impossible. (If people are worried about rampaging dinosaurs à la *Jurassic Park*, it is perfectly appropriate to explain why they cannot exist.)

The point is that only by involving the public in speculating about risks will the public feel that the attempt to define such risks is serious, not perfunctory or spurious. Precisely the same point holds regarding the mechanisms proposed to minimize the risk in question. Not only should the public be informed that, for example, potentially dangerous organisms created by genetic engineers will be "contained," they should be told what that containment means and be allowed to express their concerns about the risk management, as well as about the risk.

The second component of risk assessment is the valuational component – having judged what the risks are and ideally assessed their likelihood, it still remains to decide whether we want to take some, or all, or none of the risks we have delineated, or under what conditions we are prepared to do so. Here, again, scientific or governmental paternalism is not the appropriate tack. "Don't worry, you'll thank me later" is neither pragmatically nor ethically acceptable in a democracy. If the public is to accept any risks at all, it should be allowed to enter the dialogue on whether the risk is worth taking, not worth taking, or should be delayed in the absence of acquirable knowledge. It must be involved in determining how much weight to assign to risks and benefits, instead of this being paternalistically dictated by "experts."

My earlier analogy with an individual facing unknown risks can help clarify the point. I argued that an individual faced with the possibility of a new job at a new location is confronted with unknown risk, and that a rational person will attempt to find out as much as possible. On the basis of what he finds out, he then brings his values into play, weighs risks and unknowns against possible and actual benefits, and then makes his decision. What he chooses, of course, is still a risk – he does not have complete information and thus may have failed to consider a risk that will turn out to be pivotal. Nonetheless, he has done all he can about the risks and benefits he will confront.

Now consider this scenario: Wimpy's boss, Mr. Stalin, tells him that he is being transferred to another location. Wimpy must leave immediately; he thus cannot find out anything about the new location and can learn nothing of the benefits and risks associated with the move. Nor, of course, can he weigh them according to his values. Mr. Stalin, however, assures him that he, as the boss, has determined the risks and benefits, has also weighed them, and has concluded that Wimpy should definitely go. Now it may well be that everything Mr. Stalin says will turn out to be the case. Still and all, few of us would find such treatment acceptable. At best it is paternalism, presuming to judge what is best for another;

at worst it can be viewed as treating an individual as a piece of chattel, not as a person, totally ignoring his autonomy and dignity. In any case, such an approach is highly – paradigmatically – autocratic; if a society operated that way, we would certainly deny that it is a free society.

Paul Feyerabend has brilliantly explored the sense in which reliance on experts for making decisions for everyone is inimical to having a free and democratic society.[20] Feyerabend's point is that, in our society, government often makes policy primarily by consulting scientific experts. Such experts do not come from a value-free context, but are informed by the values underlying science, though, as we saw, they may and usually do believe that their basis is value-free. The values of individuals and other groups in society, however, may be significantly different from the values of scientists. Scientists, for example, favor science as the best way of knowing because it allows humans to predict, control, and change nature. Traditional Navajos, on the other hand, may not be very interested in predicting and changing; their value system is to live in harmony with nature, not control it. Feyerabend thus argues that, in a free democratic society, it is presumptuous to assume that science should always be the final word on social decisions, for this strategy arbitrarily favors one set of values over others.

I myself experienced an instance of Feyerabend's thesis very dramatically with a Navajo student. She had been well educated in science, holding degrees in both animal science and public health, the latter from an Ivy League school. After some years of working in public health, she had decided to return to the Navajo nation and tend sheep with her grandmother. When I asked her why, she told me the following: "I have always been interested in the fact that sheep sometimes have twin lambs. As a scientist, I learned to explain this event in physiological terms. My grandmother, on the other hand, taught me that twins are a reward for good care. I much prefer the latter explanation, so I am going back to the traditional ways."

One need not be as radical in this regard as Feyerabend to

realize that, at least in the area of genetic engineering of plants and animals in particular, the government and the scientific community in tandem have behaved toward the public like Mr. Stalin did toward Wimpy in the above example. The regulations, or lack thereof, regarding genetically engineered organisms, even the older ones governing recombinant DNA at the microbial level, have come from the scientific community and the government, with little solicitation of input from the public. Genetic engineering of animals and plants has proceeded without governmental solicitation of public participation and input; at each stage the public is presented with a fait accompli. No extensive public hearings have been held on these matters, nor have any efforts been made to engage the public in dialogue, or even to educate the public on the nature of the technology, its possible dangers as perceived by the scientific community, and its implications. It is thus not surprising that local municipalities, such as Cambridge, Massachusetts, have initiated their own debate and laws regarding genetic engineering. Even less surprising is the fact that people are highly suspicious and fearful of genetic engineering, and that surveys indicate that the overwhelming majority of people feel that they have not been, but would like to be, involved in setting policy for genetic engineering. Small wonder, then, that doomsayers and critics of genetic engineering receive great attention from the public and the media.

Let us explore this point a little further, for it is, in fact, the fulcrum for our discussion in this chapter. When one visits with biotechnologists, as I have done at numerous meetings, they are literally unanimous in bemoaning the lack of social acceptance of genetic engineering, the distrust of the field that seems so easily invoked. How can biotechnology of all sorts – not just genetic engineering of animals – they lament, be deployed for the public good in a multitude of ways, if the citizenry is still suspicious – indeed, downright fearful – of the technology?

The answer, of course, is that it cannot. The lack of public understanding, and consequent suspicion and fear, is quite understandable – all organisms greet the new and foreign with wariness, for the world contains more danger than friendliness. As I remarked earlier, things have changed a great deal since the naive 1950s faith in science as curer of all ills. In 1957, a National Opinion Research Survey indicated that 90 percent of the public agreed that the benefits of science outweighed the risks of science, and 97 percent believed that science makes life better.[21] By 1986, the 90 percent had dropped to 60 percent, according to a Louis Harris survey.[22] The same survey showed that 71 percent of the public believed that science and technology posed significant risks. The vast majority of the public surveyed supported strict controls for biotechnology and only 16 percent believed they understood DNA (the true number is probably appreciably smaller).

According to a new survey,[23] 85 percent of the public agree or strongly agree that "citizens deserve a great role in decisions about science and technology"; 82 percent feel that citizens have too little to say on deploying biotechnology; and under 23 percent have confidence in government's ability to effectively regulate biotechnology. Ninety-three percent feel that government should pay more attention to ordinary citizens on biotechnology. Fifty-three percent indicate a belief that changing organisms through biotechnology is wrong; 87 percent feel that environmental protection is more important than economic growth; and 94 percent believe that the balance of nature is easily upset by human activities. Seventy percent believe that modifying the environment for humans can cause serious problems. At the same time, interestingly enough, the survey showed that 68 percent agree or strongly agree that the government should fund more biotechnology research because of potential benefits; 67 percent disagree that only biotechnology companies will benefit from biotechnology; and 71 percent believe that they will personally benefit from biotechnology in the next five years!

The surveys echo common sense. People do see that biotechnology can benefit society and themselves personally, but they see unknown dangers in biotechnology that could harm nature and humans and generate other harms in unknown ways. They therefore wish to be involved in the decision-making process, especially as regards a new and frightening process they don't understand. When science and government fail to do this, people feel suspicious, suspect they are not being told the truth, openly distrust the government-science axis, and distrust the technology. Equally important, they are far more open to half-truths and disinformation coming from real or self-styled experts who paint a negative or apocalyptic picture of biotechnology, as I indicated in my moral version of Gresham's law.

Common sense tells us that, as individuals, in circumstances like this in ordinary life, we attempt to fill the lacunae in knowledge and communication. To take a personal example: I teach animal rights/welfare issues to animal agriculture students and science students. As mentioned earlier, I also write and lecture a great deal on these matters to audiences ranging from farmers and ranchers to medical researchers and corporate executives. When I first began to do this, I faced a great deal of prejudgment, rumors that I was crazy, radical, antiscience, antiagriculture, antihuman. I faced hostility from audiences who didn't want to hear me because they knew what I was going to say, and so on. The great temptation, of course, was to avoid such uncomfortable situations, and to talk only to like-minded people who already agreed with me. Such behavior, however, was incompatible with my mission to educate, so I did the opposite and sought opportunities to speak to those most hostile for as long as possible. I did not ignore the hostility; I attacked it head on and welcomed it, provided it was openly articulated in a rational way. Over time, I earned the trust of these diverse groups and, even more remarkably, was able to garner agreement from them on many substantive points. I found, for example, that some 90 percent of my rancher audiences believe in some sense that animals have rights, and most deplore factory farming!

The point is that ignorance and remoteness breed suspicion and hostility unnecessarily. On most moral issues, there is a broad range of potential consensus that can be achieved through dialogue. After all, we have all been brought up under the same social ethic and share most of it; if we did not, we could not live together.

The same strategy should thus be deployed vis à vis genetic engineering and other areas of biotechnology. For the technology to move forward, consensus must be achieved. For consensus to be achieved, there must be dialogue, mutual education, common vocabulary, differences and concerns clearly laid out, and mutual decision making encouraged. Yet, in point of fact, the government and the scientific community have violated all the dictates of common sense.

THE ARROGANCE OF "EXPERTS"

Consider an event that recently occurred. The FDA announced, on May 26, 1992, that food products – fruits, vegetables, and grains – altered by genetic engineering will be regulated no differently from foods created by conventional means, because they raise no new or unique safety issues.[24] The policy was developed for the FDA in cooperation with Vice President Dan Quayle's Council on Competitiveness, since the development of such new products will spur economic growth. Such products, which were engineered to be resistant to disease and insect damage, and to increase storage life and speed growth, were created by inserting genes derived from chickens into potatoes, genes derived from fireflies into corn, and genes derived from trout into catfish. Tomatoes had also been created with flounder genes. The FDA based its policy on widespread scientific belief that genetic engineering is no different in principle from what happens through traditional breeding techniques. No special labeling of foods created by genetic engineering will be mandated by the FDA.

I would argue that this incident represents a paradigmatic example of what science and government should do if they wish to assure continued public suspicion toward biotechnology and its products. As we saw in the survey, people put environmental safety (and, a fortiori, human safety) above economic growth. The motivation here was plainly economic and was dictated by a group whose name smacks of business greed, the Council on Competitiveness, headed by a person, Dan Quayle, who hardly inspires confidence and who, in any case, is not perceived as a credible expert in scientific safety. Equally important, this decision was reached with great arrogance – no public hearings were held, no attempt was made to solicit public input in any form. This in turn feeds the commonsense view that scientists and government have no regard for public concerns, have their own agenda, and are removed from ordinary common sense and its values. To announce blithely and without public discussion that tomatoes containing flounder genes are just like other tomatoes is to display either great ignorance of ordinary ways of thinking or great contempt for them, or both.

For those of us who follow biotechnology, this case inspired a sense of déjà vu. Similar insensitivity was displayed by the biotechnology industry when one of its first, and most publicized, attempts to market biotechnology products was the animal drug BST (bovine somatotropin), also known as BGH (bovine growth hormone). This product was basically designed to be injected into cows once a day during the experimental work on it, with an eye toward eventually marketing a product that would be injected every two weeks, in order to produce greater milk yield, thereby making dairy farms more efficient and capable of producing more milk. (BST works by reapportioning nutrients into milk secretion rather than fat deposition in the animal.)

The industry could not have picked a worse inaugural product if it had left the choice to Jeremy Rifkin. For BST raised public ire in virtually every conceivable segment of society. In the first place, issues of adulteration and food safety surfaced quite quickly. Despite assurances from the FDA and the biotechnology community of the safety and wholesomeness of the product, public confidence did not materialize. Other experts questioned the safety of the product, and successful consumer boycotts were initiated. In Britain, also, much publicity was garnered by those concerned about food safety.

In retrospect, the industry's willingness to press forward a product that would predictably raise questions about the safety of milk, a paradigmatic symbol of purity and wholesomeness, is astounding. It bespeaks either arrogance or stupidity. One need not be a Nostradamus to realize that parents would surely reject a product concerning which even a suspicion of danger or adulteration existed. But that was not the only folly manifest in this case. The industry should surely have realized that a public accustomed to cheap milk and to surplus milk – weaned, as it were, on pictures of farmers dumping milk to protest low prices and of warehouses full of surplus dairy products – was not likely to be overly sympathetic to a new product that increases milk yield!

Finally, BST stirred the wrath of small dairy farmers and of the multitude of citizens for whom "the family farm" is, as it were, a sacred cow. Farmers argued that if BST came into common use, small dairymen would not be able to afford its costs, would not be able to compete with large corporate dairies, and would thus be driven out of business. Advocates affirmed that BST is neutral vis à vis size of the operation.

As if all this were not enough, issues regarding the welfare of the cows to be treated with BST were also raised. It is well known that modern dairy cows, while producing significantly more milk than their predecessors, also have a somewhat shorter productive life. Whereas traditional milk cows might continue to produce for an average of three to four years, today's high producers can be culled for metabolic diseases within two to three years, and thus lose their value and get shipped to slaughter. BST, it was argued by some, would only increase this tendency. In addition, BST might also increase "production diseases," that is, those diseases associated with the stress of high productivity. BST-treated cows do, in fact, show a higher incidence of mastitis than do untreated cows.

Again, the key point, for our purposes, is that the industry was apparently blindsided by these objections. This should not have been the case. Any reasonably intelligent citizen, if informed about BST, was likely to have raised some or all of these objections. Certainly small dairy farmers would have been able to do so. And yet this didn't happen. So one can assume that the industry (i.e., those hoping to market BST) felt no need to worry about the social reception of the product – a degree of short-sightedness that was again reflected in the recent FDA decision I just mentioned. Indeed, despite the public unease about BST, the FDA approved its use in late 1993 and did so again without requiring that products from BST-treated animals carry any labels indicating it was used.

A VIABLE MODEL FOR REGULATION OF GENETIC ENGINEERING

We can now draw together the diverse threads of our discussion into a common conclusion. Given the ultimate un-assessability of risks associated with genetic engineering of animals; given the unformed nature of social values regarding genetic engineering and its risks and benefits; given the bad ethical concerns that have replaced good ones; given the definite desire on the part of the public to participate in decision making on genetic engineering; given the fact that people believe that biotechnology will benefit them, yet also are ignorant and fearful of it – all these vectors militate in favor of meaningful, serious regulation that addresses these issues, for plainly biotechnology will stand or fall with public acceptance or rejection, not with the progress of the science. And the nature of that regulation seems to me patent: If the public is to feel protected, safe, and involved, it must be an active participant in the acceptance and rejection and monitoring of genetic engineering.

For these reasons, I would propose that the following sort of regulatory structure serve as a model for achieving a win-win situation regarding biotechnology. In the first place, the government should hold an extensive series of public hearings, all over the country, on genetic engineering of animals (or on various aspects of biotechnology). At these meetings, scientists and industry representatives who are honest, knowledgeable, and capable of communicating with the general public should be in attendance, as should responsible, knowledgeable critics of biotechnology. After a variety of such meetings, a list of major concerns about the technology should be drawn up – concerns about risks of all sorts: risks to humans, to the environment, to social institutions, and to the animals. (I shall discuss the last in the next chapter.) Absolutely crucial to this process is the notion that no risks or concerns should be dismissed out of hand, especially by the experts. Such a list should then be accompanied by brief, intelligible discussions of the ways in which these putative risks could be managed or minimized and of the costs associated with such management. At the same time, the realistic, putative benefits should be outlined. The resulting document, drafted in lay language (not federal bureaucratese) by a carefully chosen, consummately skillful writer, knowledgeable in the relevant science and sensitive to social concerns, should be circulated widely.

When this process is complete, and, in essence, a simple map of the logical geography of the issues is in common circulation, the second phase of regulation would commence. This would involve federally mandating broadly representative local committees to judge and pass on proposals for genetic engineering of animals (or whatever other biotechnology is of concern). Such committees would be of manageable size, say about a dozen (jury size), and would consist of people broadly representative of community concerns – a family physician, a teacher, a farmer, a blue-collar worker, a white-collar worker, and so on. The role of the committee would be twofold – first of all, to hold public hearings on proposals for genetic engineering of animals to be done in the community (primarily an educational and learning process), guided by but not restricted to, the concerns outlined in the document I proposed. Second, the role of the committee would be to weigh the results of the hearing according to certain preestablished general criteria (e.g., risk/benefit ratios, potential suffering of animals, etc.) and, ultimately, to accept or reject the project. The group could solicit additional testimony from the industry or from critics. (Obviously, there are many logistical problems to be solved. For example, what does one do in an area where there are vast numbers of projects; how does one minimize corruptibility of the committee, etc. But we are here interested in the concept, not the details.) Monitoring the project would be accomplished by a newly founded federal agency.

The advantages of such a system are clear. The industry would know what hoops it needed to jump through. The public would be significantly involved in the control of genetic engineering; the industry would lose the image of being shadowy and remote. Most important, the public would become educated about genetic engineering, and the fear and suspicion born of ignorance would be dispelled, while legitimate concerns would become focused. Only in this way can the general public feel that it has control of and identification with the extraordinarily powerful new technology that it currently cannot cognitively process intelligently.

Again, there·are problems of inconsistency of committees in different areas (though this will probably eventually homogenize as more and more experience is gained – we have seen this occur in animal care committees that pass on animal research); fear of companies about revealing things that give them an edge; time lost to deliberations. Nonetheless, I believe that such a process is the most fair, expeditious, and judicious way to generate social identification with biotechnology. In the end, though certain approaches to genetic engineering will probably be lost unfairly, I believe that common sense will by and large make the right decisions. Irrational concerns will be dispelled by open dialogue and education; genuine concerns will be sharpened and pointed. And most important, the process will be as democratic as it can be so that society as a whole is more likely to have a stake in the results of the discussion.

Cynics among my readers are probably derisively gagging at what they undoubtedly perceive as my Pollyanna-ish, utopian, *Sound of Music*, Mickey Mouse alleged solution: "Why, ordinary people can't spell biotechnology, let alone understand it, let alone pass judgment on it!" Such responses are quite common in my experience, especially from academics.

Yet we would all do well, in our antidemocratic moments, to recall that contempt for the public is ill founded, since, as both Pogo and Heidegger remarked in another context, *"They* is *us."* In other words, consider the alternative to trust in the democratic public; it is trust in an individual or a subgroup of the public that, historically, offers no greater comfort. In actual fact, as I remarked earlier, those of us like myself who interact with the public on difficult and novel issues, and who take the trouble to explain rather than patronize, find ourselves rewarded by the response of the vast majority of people, who will expend great effort and attention to understand and to be fair. In any event I see no more plausible alternative for defusing public suspicion of biotechnology and for assuring democratic consideration of concerns by the population that will be affected.

There is another possible objection to the viability of our approach to public participation in decision making regarding genetic engineering. This is the view that can be succintly expressed as a heartfelt "Oh God! Not another law chartering another federal bureaucracy!" Why not, the argument continues, let the market rule? If people are not convinced of the safety or the value of some product resulting from genetic engineering, they can vote with their pocketbooks and simply not buy the product in question, as consumers have already done with BST and milk. The onus would then be on genetic engineering companies to convince the public that their products were safe and worth buying, and the government would be kept out of it.

Such an objection can best be answered by considering each of its components separately. In the first place, let us consider the claim that we are proliferating regulation in a society that is already overregulating all aspects of life and correlatively proliferating bureaucracy. One may well believe that we are generally overregulated in many areas and also believe that in other crucial areas, we are underregulated or lacking a reasonable articulated social policy.

As I shall discuss in detail later, such an area was exemplified by the use of animals in research. By the early 1980s, it was clear that the public entertained a fairly high level of suspicion about the proper treatment of such animals. A series of well-publicized atrocities further deepened that suspicion, and some mechanism was needed to assure, or reassure, the public that proper treatment was required, not voluntary, and to serve notice to researchers that provision of good care to animals was a serious concern of public morality. At the same time, no thinking people wished to erect meaningless bureaucratic barriers in the process of scientific activity.

Those of us who wrote what became federal law in 1985 realized that what I have called scientific ideology blinded scientists to some of the fundamental moral dimensions of animal research, specifically in that ideology's disavowal of ethics as relevant to science and its professed agnosticism regarding animal consciousness, pain, and suffering (which I will discuss fully in the next chapter). So our approach to law was to create a system of enforced self-regulation, wherein scientists would be compelled by the law to deal with what they had hitherto ignored, recognizing and controlling animal pain and suffering. This was to be accomplished by local committees consisting of both scientists and nonscientists evaluating research proposals in terms of some general criteria, for example, is adequate provision made for control of pain and suffering? This approach assured that concern for animal pain would enter into scientific deliberations. We believed that, eventually, such considerations would become second nature to scientists and thus that the law was essentially a vehicle for introducing an educational vector into the insular scientific community. At the same time, public participation on these committees would also help educate the public to the fact that scientists were dealing with these issues.

In other words, the law became a catalyst for broadening thinking and discussion rather than a series of bureaucratic intrusions. Increasing numbers of scientists began to think about what they had hitherto ignored and to educate their students to the social moral concerns that had motivated the law in the first place. It is now generally acknowledged even by those who once opposed the passage of the law that its effect on scientists has been salubrious. At the same time, public suspicion of the research process has been reduced.

I see the same sort of role for law in the case of genetic engineering. If anything, the lack of common ground is even more dramatic here. The scientists (and companies) involved in this technology clearly do not understand public concerns – this much has been shown by the BST case and the FDA action (or lack of action) discussed earlier. At the same time, the public has virtually no understanding of biotechnology. A law that mandates dialogue and deliberation, such as I described above, essentially guarantees (as much as anything can) that the level of sophistication of all parties will rise. Biotechnologists will hear public concerns in an immediate way; the public will be in a position to have its questions answered directly. All parties to what is currently blindman's buff will be playing on the same court with the same rules.

In other words, it is wrong to see this proposal as creating more bean counters. Our law would create a mandated context for dialogue, with a minimal role for bureaucrats and bean counters – indeed, their major significant role would be to assure that the dialectical process was indeed respected.

Even if I am correct in my view of law as providing for compulsory education, the second part of the objection still remains: Why not leave this whole process to the companies marketing biotechnology? The answer is simple – what would occur would most likely amount to propaganda; slick, manipulative advertising that, like all advertising, does not truly educate, but persuades. Thus the public would come no closer to understanding what is very likely the most monumental technological revolution in human history. Furthermore, the door would be open for opponents of biotechnology to adopt the same tactics, resulting in a series of media assaults on the public mind reminiscent of – and with all the educational value of – a presidential campaign. Overstatement, deception, exaggeration, and appeal to authority would rule. Furthermore, local concerns would almost certainly not be expressed or dealt with. Even if a lucrative East Coast market is created through advertising for a biotechnology product based in Wyoming, there is no assurance that the people in Wyoming would be comfortable with the research or production requisite to marketing the product.

Further, I believe that the process I describe is actually fairer to the companies engaging in biotechnology. Recently, the Campbell company put a good deal of research money into creating a genetically engineered tomato, only to withdraw the product because it appeared that the public would not accept it. The process I described, which involves the public and considers public response from the very beginning of creating a product, would seem to assure constant feedback to a company before it poured too much money down the drain. If our procedure had been followed in the BST case, the company involved could have had a good indication of public concerns early, rather than late, in the game, and it could have either addressed them or truncated the project before excessive amounts of money were uselessly expended.

In sum, I would argue that a minimally bureaucratic law of the sort I described provides the best mechanism for both managing the risks of biotechnology and, even more importantly, helping the public to assimilate and shape this new technology of unprecedented power.

POSSIBLE DANGERS: EVOLUTION IN THE FAST LANE

What are the sorts of concerns that are likely to appear relative to risk and danger associated with genetic engineering of animals? I will present those that appear significant to me on the basis of both extensive reflection and extensive dialogue. My discussion is not intended to be exhaustive, and I suspect items will be both added and deleted as our knowledge and understanding grow. But some such discussion is necessary in order to at least prime the dialectical pump to encourage rational dialogue about the risks associated with genetic engineering of animals.

One of the primary concerns about genetic engineering

grows out of the rapidity with which such activity can introduce wholesale changes into organisms. The biotechnology community and the FDA have lately taken the position regarding genetic engineering that what is important to look at in terms of safety and danger is the *product* of a piece of genetic engineering, not the *process*.[25] In other words, whether we create changes in animals and plants by way of traditional genetic engineering, namely selective breeding for desired traits, or by way of the tools of the new technology, the issues that emerge are the same: Is the resultant product dangerous or diseased or problematic in some other way? The mechanism by which one induces the changes, it is argued, is fundamentally irrelevant.

The point is legitimate as far as it goes. It is also the case, however, that there might be something about the process (genetic engineering) that does dispose what is created to be more dangerous than what is produced by more traditional breeding. I believe there is a difference between the processes, which makes genetic engineering bear watching more closely than traditional breeding – the rapidity with which one can change organisms that I mentioned earlier. Traditional genetic engineering was done by selective breeding over long periods of time during which one had ample opportunity to observe the untoward effects of one's narrow selection for isolated characteristics. But with the new techniques of genetic engineering, we are doing our selection "in the fast lane," and thus we may not detect the problematic aspects of what we are doing until after the organism has been widely disseminated.

Another way to put the same point is that with traditional breeding, there is an enforced waiting period necessarily associated with attempting to incorporate traits into organisms. In the animal area, especially, one can significantly change animals from the parent stock, but it will take many generations to do so, during which time one has ample opportunity to detect problems with the genome one is creating, or with its phenotypic expression. To be sure, as occurred with the breeding of many purebred dogs, one may choose to disregard the untoward effects; hence there are literally hundreds of genetic diseases associated with purebred dogs – bleeding diseases in Dobermans, cancer in boxers, breathing problems in bulldogs, skin problems in shar-peis, and so forth.[26] But the point is, one could see the problems developing if one cared to do so. With genetic engineering, however, one can insert the desired gene in one effort, and the problems that emerge may be totally unexpected.

There are many instances of this, in fact, even in traditional breeding. One famous example of this concerns corn and grows out of the phenomenon known as *pleiotropy*, which means that one gene and its products control or code for more than a single trait. In this case, breeders were interested in a gene that controlled male sterility in corn, so that one could produce hybrid seeds without detasseling the corn by hand, which is very labor intensive. So the gene was introduced in order to provide genetic detasseling. Unfortunately, the gene was also responsible for increased susceptibility to southern corn blight, a fact no one was aware of. The corn was widely adopted, and in one year a large part of the corn crop was devastated by the disease.

Similarly, when wheat was bred for resistance to a disease called blast, that characteristic was looked at in isolation and was encoded into the organism. The backup gene for general resistance, however, was ignored. As a result, the new organism was very susceptible to all sorts of viruses that, in one generation, mutated sufficiently to devastate the crop.

What we have then, vis à vis the danger associated with genetic engineering, is what philosophers call an a fortiori situation. If such unanticipated consequences can and do occur with traditional breeding, where one of necessity proceeds slowly, how much the more so does the danger of unanticipated consequences loom when one is creating transgenic animals? When one inserts a sequence of DNA (a gene) into an organism, one cannot anticipate pleiotropic activities, where the gene affects other traits one has not anticipated. By the same token, one may have overlooked the need for more than one gene to get the desired result phenotypically. Any of these factors can produce a variety of conditions deleterious to the organism.

The way to control this risk, then, whether one is doing traditional breeding or taking transgenic shortcuts, is to do a great deal of small-scale testing before one releases or depends on the new organism. If one is inserting genes derived from other species, the unexpected results and possible adverse effects are even less predictable and probably more likely. (Note that I am here assessing "adverse effects" as those that humans genetically engineering the organism find undesirable, e.g., susceptibility to a disease. In the next chapter, I will deal with the more difficult issue of the situation in which humans may wish to create an effect that harms the animal or do not care about an effect that harms the animal.)

The second type of danger resulting from fast-lane genetic engineering of animals can be illustrated by reference to food animals. Here the isolated characteristic being engineered into the organism may have unsuspected harmful consequences to humans who consume the resultant animal. Thus, for example, one can imagine genetically engineering faster growth in beef cattle in such a way as to increase certain levels of hormones that, when increased in concentration, turn out to be carcinogens for human beings over a thirty-year period, or teratogens, like diethylstilbestrol. The deep issue here is that one can of course genetically engineer traits in animals without a full understanding of the mechanisms involved in phenotypic expression of the traits, with resulting disaster. Ideally, though this is probably not possible either in breeding or creating transgenics, one can mitigate this sort of danger by being extremely cautious in one's engineering until one has at least a reasonable grasp of the physiological mechanisms affected by insertion of a given gene.

POSSIBLE DANGERS: NARROWING OF THE GENE POOL

A third general kind of risk growing out of genetic engineering replicates and amplifies problems already inherent in selection by breeding, namely the narrowing of a gene pool, the tendency toward creation of genetic uniformity, the emergence of harmful recessives, the loss of hybrid vigor, and, of course, the greater susceptibility of organisms to devastation by pathogens, as has been shown to be the case in crops.

So, once again, we encounter a problem that already exists in traditional breeding. As we find the traits we consider desirable, we try to incorporate these traits into the organisms we raise, be they plant or animal. We continue to refine and propagate these animals and plants until a particular genome dominates our agriculture. In other words, we put all our eggs in one basket. The number of strains of chicken in production of eggs and broilers, for example, has decreased precipitously since the rise of large corporate domination of the industry during the last forty years. What this means in practical terms is that the industry stands and falls by what it considers the few superior genomes it has developed. If circumstances change, or if a new pathogen is encountered, wholesale devastation of the population will of necessity occur and has occurred, for example by Newcastle disease or influenza. Loss of genetic diversity means loss of potential for adaptation to new circumstances.

The way in which genetic engineering can accelerate this tendency is clear. Suppose a "superior" animal is created transgenically with great rapidity. Those who utilize this animal gain a clear competitive edge, be it because of increased disease resistance, greater efficiency in feed conversion, greater productivity, or whatever. In order to compete, other farmers replace their stock with this animal, as old strains are viewed as obsolete. The entire branch of agriculture becomes a monoculture, with the extant gene pool severely limited. Over a period of time, however, an untoward characteristic of the new genome emerges; be it disease susceptibility, reproductive problems, stress susceptibility, or whatever. A potentially disastrous situation forms because the potential for responding to the crisis has been lost with the loss of genetic diversity. Alternatively, social or economic circumstances may change so as to require change in agricultural practices or locale such that the extant genome does not fit well with the new circumstances. Once again, the presence of a mono-cultural animal militates against the sort of quick, reasonable, and efficacious response that a diverse gene pool would provide.

In the end then, genetic engineering of animals runs the risk of accelerating the tendency that is already established in at least certain portions of animal agriculture – the chicken and egg industry, for example. Interestingly enough, my colleague Dr. George Seidel[27] informs me that the poultry industry is in fact reluctant to employ genetic engineering, for in the time it takes to introduce a new gene construct, say a gene for disease resistance, into the whole population, the industry can make vast progress in efficiency by traditional breeding, say for fast growth, which essentially means more profits. While one would be attempting to propagate the new gene introduced transgenically by breeding it into the population, one could not at the same time continue to select for the other economically desirable traits, and the trade-off is simply not worth it economically. In other words, the poultry industry would rather not slow down its highly efficient breeding program even to introduce a worthwhile gene transgenically. This means that transgenic technology is unlikely to invade the poultry industry for some time. But, this exception notwithstanding, my general concern still stands!

On the other hand, genetics can have an opposite effect, by preventing the complete loss of genetic diversity in an age of monoculture. By establishing a "gene library" of animals whose genome is not considered of sufficient value to propagate (i.e., a collection of germ plasm), genetic engineering can help to preserve strains that might otherwise disappear. (This is currently done to some extent by rare-breed fanciers, who breed such animals, but could be rendered far more effective by the new technology.) Furthermore, genetic engineering could help to widen the gene pool that is available to animal breeders, in the same way that artificial insemination technology made a great deal more new genetic material available to beef breeders.

POSSIBLE DANGERS: UNWITTINGLY SELECTING
FOR PATHOGENS

A different area of concern about potential danger associated with transgenic animals arises out of the fact that when one changes animals, one can thereby change the pathogens to which they are host. Such a scenario could arise in a number of different ways. In the first place, if one were genetically engineering the animal in question for disease resistance to a given pathogen, one could thereby unwittingly select for new variations among the natural mutations of that microbe to which the modified animal would not be resistant. This new organism could then be infectious to these animals, other animals, or humans. In other words, such genetic engineering could, in essence, become a selectional pressure for changing the population of pathogens hosted by the organism.

This is not science fiction. This same sort of thing has occurred, as mentioned earlier, by virtue of widespread use of antibiotics in humans and animals, be it for therapy or for growth promotion in farm animals. Most of us have heard of the battle between pathogen change and antibiotic development. We use a given antibiotic to fight a given disease, say streptococcal infection. We effectively wipe out the susceptible strain, thereby leaving, as it were, a clear field for another strain not susceptible to that antibiotic. That strain proliferates, while drug companies modify the antibiotic in some manner so that it can engage the newly flourishing pathogen. The latter is effectively destroyed, and the cycle begins again. A similar problem has recently arisen with drug-resistant tuberculosis microbes. The same sort of thing could presumably occur with any form of disease mitigation, including genetic resistance. Once again, though, the possible danger resulting from genetic engineering is no different in principle from what we have already encountered in a different context.

Another way in which genetic engineering of animals could affect pathogens might occur even when one was not selecting for disease resistance. Suppose one were genetically engineering farm animals to increase rumen efficiency. A corollary of such a change might be to change rumenal temperature, or pH, or some other variable. By so doing, one changes the microenvironment in which various microbes dwell. Such a change in that environment could very well provide selection pressures to change the nature of the population of these microbes. This, in turn, might affect the pathogenicity of the microorganisms that inhabit the animal in unknown and unpredictable ways. And the more precipitous the genetic change, the more inestimable the effects on the pathogens are likely to be. The scenario we have outlined, of course, is not restricted to a rumenal change, but is in principle possible with any genetic change that will have effects on the microenvironment in which pathogens flourish.

POSSIBLE DANGERS: GENETICALLY ENGINEERED DISEASE MODELS

A fifth way in which genetic engineering of animals could produce significant risks does not concern farm animals, but animals genetically engineered to serve as "models" in biomedical research. (I will have a great deal to say about some vexatious ethical questions associated with such animals in the next chapter, but here I will focus on potential risk growing out of the creation of such animals.)

During the late 1980s, a mouse model for AIDS was created at the National Institute of Allergy and Infectious Disease of the National Institutes of Health (NIH).[28] HIV–1, the pathogen that causes human AIDS, will naturally infect only humans and chimpanzees, and the chimpanzees do not in fact develop any symptoms of the disease. Researchers were thus interested in creating an animal "model" for the disease to enable them to study both the course of the disease and possible therapies against it. To do so, the researchers introduced the AIDS genome into mouse embryos by microinjection and then propagated these animals by breeding. Some of the progeny developed AIDS-like symptoms, and their tissues produced infectious HIV particles. Alternatively, the so-called SCID (severe combined immunodeficiency) mouse was created by genetic engineering to model AIDS.[29] These animals have a genetic defect that wipes out the animal's own immune system. The animals are then given the genes for a human immune system and infected with AIDS.

There are two clear sorts of risks associated with the development of such animals. The first one is obvious and is associated with the nightmarish possibility of infectious AIDS mice loose in the general population. The NIH believed that the benefits growing out of having such an inexpensive model for AIDS justified the potential danger, and it took what one scientist derisively called "overkill" measures to minimize the risks. All of the animals in the NIH model were kept in a stainless steel glove box closed system that was itself kept in a facility classified as having the highest level of biosafety and maximum microbial containment recognized by the scientific community – BL4, or Biosafety Level 4. Biosafety Level 4, which mandates all personnel wearing what are essentially separately ventilated "space suits," is generally used for animals harboring the most dangerous, exotic, and life-threatening pathogens. To give the reader a basis for comparison, anthrax and plague, hardly pussycats among diseases, are kept at BL2. Examples of pathogens kept at BL4 are the nightmarish Lassa fever and Marburg virus. At any rate, NIH containment of the AIDS mouse went beyond BL4. The containment boxes in the BL4 facility were surrounded by bleach-filled moats and by traps designed to kill the mice were they to escape from the box. Such extraordinary caution is, I believe, both wise and prudent in the face of our earlier discussions.

But yet another danger emerged from the creation of these animals. In 1990, it was announced in *Science* that the AIDS virus inserted into the mice described above can interact with a common mouse virus to produce what is essentially a new pathogen in a synergistic reaction.[30] The new virus, or the modified AIDS virus, can acquire the ability to reproduce far more rapidly than it normally does and also to infect new kinds of cells. These viral variants might also spread by new routes, for example, by transmission through the air, which the ordinary AIDS virus cannot do. Not only does this synergy with indigenous viruses create significant additional risks, it also vitiates the fidelity of the animal model, since the modified virus is no longer a good model for HIV–1.

Like the risk of animal escape, such risks can be minimized by extraordinary containment policies. What the current example bespeaks, however, is the need for not being cavalier about what is initially thought to be very unlikely, but to treat any potential risk, however far-fetched it may seem to the state-of-the-art science, as a serious possibility. When one is genetically engineering animals, what we can do far exceeds what we can know and predict, and it is thus good policy to prepare for the worst case.

POSSIBLE DANGERS: ENVIRONMENTAL

A sixth category of risks is environmental and ecological and grows out of genetically altering animals and either their getting loose in an environment other than the one for which they were intended, or their being released into an environment for which they were intended, but with negative, unanticipated consequences. This is probably a greater concern for genetically engineered plants than for genetically engineered animals, as plant propagation and dissemination of germ plasm is, generally, far easier, quicker, and more widespread than is animal (as we all know from the rapid and widespread appearance of weeds in our lawns and pastures). But the problem is real for animals as well.

Ecosystems are very much like automobile windshields – they are built to withstand a great many assaults with little or no apparent damages; hit them just right (or wrong) sometimes in a minor, apparently trivial way, however, and they can sustain major structural damage. Consider our first sort of concern: the escape of genetically engineered organisms beyond their intended confines. An excellent possible example of that is provided by the aforementioned AIDS mouse; were it somehow to get out of the BL4 facility and begin to interbreed with wild mice, the viral genome could (and probably would) get into the general mouse population. There it could in turn experience the sort of synergistic transmutation I mentioned earlier by interacting with indigenous mouse viruses. This could result in massive infection of a variety of animals (and humans), with wholesale and unpredictable devastation to the ecosystem as a whole. While there has been, to my knowledge, no escape of transgenic animals from the limited facilities for which they were intended, enough examples exist of nontransgenic animals getting out of intended environments and into unintended environments, that we should be quite sensitive to the potential disasters that could be occasioned.

I have already mentioned killer bees; that is an example of an escape that is still not under control and whose consequences are still unclear. Another current example is the zebra mussel, which found its way from Europe by attaching itself to ships and has colonized waterways in the Great Lakes. It is clogging water intake pipes at filtration plants, competing with native species, and generally wreaking havoc, and again no one is clear on how to reverse its entrenchment. A similar problem has occurred in Florida, where a giant land snail, apparently imported as an oddity by fanciers, is causing great harm in Florida waterways. Tree snakes, accidentally released in Guam, have decimated bird populations otherwise impervious to predators. And one could recite many similar examples.

What this sort of case points to is the need for the same sort of "overkill" I mentioned in relation to the AIDS mouse. In Crichton's *Jurassic Park*, he talks precisely of such barriers to escape and presents plausible accounts of how they are nonetheless circumvented. Indeed, Crichton describes not only physical barriers but biological barriers. For example, the animals can be engineered to be dependent on elements of their diet that can only be supplied in specially formulated feed – if they escape and fail to receive this vital dietary component, they die. In other words, part of the genetic engineering involves creating dependence upon the restricted environment. Crichton develops scenarios whereby these barriers can be breached as well, but clearly a proliferation of types of safeguards, as well as intensification of degrees, is a plausible approach to risk minimization where the benefit of creating the organism is patently significant enough to warrant the sorts of risks involved.

One cannot minimize the dangers associated with the accidental release of organisms; thus one should do everything possible to assure it does not occur. Even more complex, however, is the issue of the dangers associated with deliberate release of organisms, where thinking and planning are done in advance, and the addition of a new creature to an ecosystem is intentional. Again, there are numerous examples, not involving genetic engineering, of the problems that can arise, even when it is believed that one has properly thought the matter through in advance. The classical example of such ecological disaster occurred in Hawaii, where the mongoose, originally an Asian animal from the Indian subcontinent, was imported and released to control rodents damaging the sugar cane crop. The result caused ecological disaster, as yet unresolved; the mongoose devours not only noxious rodents, but a wide variety of the indigenous fauna. Another famous example is provided by the introduction of the European rabbit into Australia; in the absence of natural predators, the animal bred, as it were, like a rabbit, devastating forage. This turn led to extraordinary measures being introduced in an attempt to control the animal, such as introducing rabbit diseases. Still another example comes from introducing the arctic fox, native to the Arctic mainland, onto some Arctic islands, where it decimated bird populations.

There is an a fortiori argument implicit here: If we can take animals whose characteristics are well-known, well understood, and reasonably predictable and put them into environments that are familiar, and we still occasion disaster – sometimes disaster that we can't reverse – how much more likely are we to do so with new organisms, whose traits we do not yet understand? Recall that some genes control multiple traits. It may well be that some gene inserted to yield a desirable trait may be found to have deleterious implications that are expressed only in the wild, not in the lab. More generally, we do not know what the new animal is likely to affect regarding the ecosystem as a whole. Even disregarding chaos notions, ecological systems are not rows of Newtonian dominoes – causation is multidimensional and multifaceted.

To their credit, certain components of the scientific community have been aware of the potential dangers associated with release into the environment of genetically engineered organisms. Most fittingly, the Ecological Society of America addressed this question in 1989 in a special report authored by a group of ecologists.[31] In addition, the journal *Fisheries* published a series of articles on the same subject.[32]

What, then, are the potential dangers, or sorts of dangers, associated with environmental release of genetically engineered animals? Some of them will be similar to the cases I cited above, where deliberate or accidental introduction of novel animals caused unsuspected harm. Others might be more subtle or extreme. Obviously, when we are dealing with intensively maintained animals, for example, swine in confinement systems, such dangers are reduced significantly, since, generally speaking, the animals are to a large extent isolated from interacting with any environment beyond the artificial one in which they are maintained. This is not, of course, to suggest that these systems or artificial environments do not affect the larger environment – waste disposal, methane accumulation, water use and contamination are also major ecological impacts of confinement agriculture. But one need not worry in these cases about the animals themselves having a direct impact. There is relatively little danger of genetically engineered swine kept in confinement taking over the countryside, for example. As I remarked earlier, however, they may well host new pathogens that could move beyond the artificial environment.

On the other hand, there are genetically engineered animals that would be (or in fact have been) released into the broader environment. These include predator insects designed to perform biological pest control by preying on noxious insects; genetically engineered fish designed to be cold resistant, or disease resistant, or to grow faster; and so on.

The more any organism differs from its parent stock, the more difficult it will be to predict its effect on the environment if it is introduced therein. One could be creating new pests, if the new animal is too successful. One could thereby be crowding out other species, plant and animal, one did not wish to affect. One could conceivably potentiate current pests by crowding out their natural enemies, or, in the case of plants, put the new selectively advantageous genes into the weed population by crossbreeding. The new organisms could display traits in an environmental setting that were not evident in laboratory or other controlled situations, which traits had unanticipated effects on the biotic community. (Recall my discussion of pleiotropy.) The introduction of predator insects could essentially serve to select for insects not vulnerable to such predators. In other words, the introduction of superpredators of any sort could help generate "superprey." The products of genetically engineered organisms could prove unpredictably problematic to other organisms. For example, genetically altered beef cattle could end up producing feces or urine that might harm plants or soil microbes. Heavier cattle might compact soil.

Many of these sorts of concerns are discussed in the Ecological Society report mentioned above, albeit largely with regard to plants and microbes, rather than animals. The principles, however, apply universally. As the report states: "Although the capability to produce precise genetic alterations increases confidence that unintended changes in the genome have not occurred, precise genetic characterization does not ensure that all ecologically important aspects of the phenotype can be predicted for the environments into which an organism will be released."[33] The report also recognizes that, in general, introduction of new genes into organisms will probably reduce fitness of the organism in an uncontrolled environment, because new and additional genes exact greater metabolic costs and because normal physiological processes can be disrupted in unpredictable ways, as I discussed earlier.[34] On the other hand, there are known exceptions to this generalization, where new gene introduction seems to be neutral or even to enhance competitive fitness. One cannot predict in advance which of these options will in fact occur in a given case. In addition, the report points out that released organisms will be subject to natural selection forces and thus will change over time in unpredictable ways as they evolve. This, in turn, will place new selection pressures on other organisms, again with unpredictable results. Crichton's *Jurassic Park* again explores these concepts in fictional but plausible ways.

The report also warns that "the absence of an immediate negative effect does not ensure that no effect will ever occur."[35] For all of the above reasons, the report commonsensically suggests that, prior to general release of genetically engineered organisms, one should first begin with laboratory studies, then move to small-scale field tests under controlled conditions that minimize the possibility of spread, and that all this take place within the context of an adequate regulatory structure. Current regulatory practice is inadequate, in that, whereas universities and other federally funded institutions must review all genetic engineering projects, commercial research, which constitutes the bulk of genetic engineering, is exempt.

As I emphasized in my earlier discussion of a local, democratic regulatory structure, the report stresses the need for a case by case examination of each genetic engineering project, rather than the use of mechanical, inflexible rules. It also offers a table of reasonable, commonsensical general principles to help guide thinking, which I have reproduced as Table 2.1.

The table is meant not as an algorithm, but as a guide for vectoring relevant ecological considerations into the judgment of risk. In addition, the report stresses the need to maintain vigilance and the need to avoid falling into a "we haven't had any problems with genetic engineering; therefore we won't" mind-set. "The absence of problems at an early stage suggests that the screening mechanisms are working correctly, but should not be interpreted to mean that the introduction of all genetically engineered organisms is inherently safe."[36] In addition, there ideally needs to be international cooperation to avoid exporting of organisms shown to be safe in one ecosystem to others where the testing may not be valid.

Everything in this reasonable, commonsensical, prudent document is compatible with my earlier suggestions. All of the suggestions offered in it could be implemented by the system I described above. General guidelines could be federally promulgated and utilized as a framework by local groups. Subsequent project oversight could be delegated either to existing federal agencies or, better yet, to specially trained agents of a new agency. Though we all resist bureaucratic proliferation, we should also realize that just because there is too much regulation in many areas does not mean that there is enough in others.

The Ecological Society's recommendations are nicely complemented by a series of articles in *Fisheries*. Fish are the animals that have hitherto been most subject to genetic engineering, and they have the greatest potential for imminent utilization both under controlled conditions (aquaculture) and for release into natural environments. While the Ecological Society's remarks were largely based on issues associated with plant and microbial release, these papers assess environmental risk and risk management vis à vis animals.

The three articles, written by Anne R. Kapuscinski and Eric M. Hallerman, explore various aspects of transgenic animal creation, as indicated by their titles: "Transgenic Fish and Public Policy: Anticipating Environmental Impacts of Transgenic Fish"; "Transgenic Fish and Public Policy: Regulatory Concerns"; "Transgenic Fish and Public Policy: Patenting of Transgenic Fish."[37] I will summarize some of the major salient points as they relate to our discussion.

The authors point out that, although no environmental release of transgenic fish has yet occurred, a variety of projects are under way. These include genetically engineering fish for growth, increased efficiency of food conversion, changes in tolerance for cold, tolerance for salinity, tolerance for pH variations, disease resistance, and drug resistance. Other changes besides those specifically designed would also be likely to occur, due to pleiotropy; unintended affected behaviors could include alterations in seasonal migration, reproduction, prey selection, territoriality, and habitat selection.

Table 2.1. Attributes of organisms and environments for possible considerations in risk evaluation*

A. Attributes of genetic alteration

	Level of possible scientific consideration		
	Less ⟶		⟶ More
Characterization	Fully characterized		Poorly characterized or unknown
Genetic stability of alteration	High (e.g., chromosomal)		Low (e.g., extrachromosomal)
Nature of alteration	Gene deletions (unless host range altered)	Single gene added	Multiple genes added
Function	None (no expression or regulation)	Regulation of existing gene product	Synthesis of gene product new to parent organism
Source of insertion	Same species	Closely related species	Unrelated species
Vector	None	Non-self-transmissible	Self-transmissible

	Less ⟵ Level of possible scientific consideration ⟶ More		
Source or vector	Same species; nonpathogen	Closely related species; nonpathogen	Unrelated species or pathogen
Vector DNA/RNA in altered genome	Absent	Present, but nonfunctional	Functional

B. Attributes of parent (wild type) organism

	Less ⟵ Level of possible scientific consideration ⟶ More		
Level of domestication	Unable to reproduce without human aid	Semi-domesticated; wild or feral populations known	Self-propagating, wild
Ease of subsequent control	Control agents known		No known control agents
Origin		Indigenous	Exotic
Habit	Free-Living		Pathogenic, parasitic, or symbiotic

(continued)

Table 2.1. (cont.)

	Relatives not pests	Relatives pests	Pest itself
Pest status			
Survival under adverse conditions	Short term		Long term (e.g., spores, cysts, seeds, dormancy)
Geographic range, range of habitats	Narrow		Broad or unknown
Prevalence of gene exchange in natural populations	None		Frequent

C. *Phenotypic attributes of engineered organism in comparison with parent organism*

Level of possible scientific consideration

	Less		More
Fitness	Reduced irreversibly	Reduced reversibly	Increased
Inffectivity, virulence, pathogenicity,	Reduced irreversibly	Reduced reversibly	Increased

Host range	Unchanged		Shifted or broadened
Substrate, resource	Unchanged	Altered	Expanded
Environmental limits to growth or reproduction (habitat, microhabitat)	Narrowed but not shifted		Broadened or shifted
Resistance to diseased, parasitism, herbivory, or predation	Decreased	Unchanged	Increased
Susceptibility to control by antibiotics or biocides, by absence of substrate, or by mechanical means	Increased	Unchanged	Decreased
Expression of trait	Independent of environmental context		Dependent on environmental context
Similarity to phenotypes previously used safely	Identical	Similar	Dissimilar

(continued)

Table 2.1. (cont.)

D. Attributes of the environment

	Level of possible scientific consideration		
	Less		More
Selection pressure for the engineered trait	Absent		Present
Wild, weedy, or feral relatives within dispersal capability of organism or its genes	Absent		Present
Vectors or agents or dissemination or dispersal (mites, insects, humans, machines, wind, water, etc.)	Absent or controllable		Present, uncontrollable
Direct involvement in basic ecosystem processes (e.g., nutrient cycling)	Not involved	Marginally involved	Key species

Characteristic			
Alternative hosts (partners), if organism is involved in symbiosis (mutualism)	Absent		Present
Range of environments for testing or use; potential geographical range	Very restricted		Broad, widespread
Simulation of test conditions	Not difficult to simulate realistically		Very difficult to simulate realistically
Public access to test site	Tightly controlled	Limited	Uncontrolled
Effectiveness of monitoring and mitigation plans	Proven effective		Untested or unlikely to be effective

*Position on scale is only qualitative or semiquantitative. The importance of position on one scale may be contingent on another scale. The importance of particular scales will vary with different cases.
Source: J. M. Tiedje et al., "The Planned Introduction of Genetically Engineered Organisms: Ecological Considerations and Recommendations," Ecology, vol. 70, no. 2 (1989), pp. 308–10. Reprinted with permission from the Ecology Society.

The release of a large number of altered fish could radically change the species' ecological niche, with major effects possible on other elements of the ecosystem. The entire food web could be restructured or destabilized, for example.

The authors describe a continuum of risks associated with the environment into which the transgenic animals are placed. Clearly, the lowest risks would be associated with indoor, research laboratory environments (as with the AIDS mouse in BL4 containment). On the other end of the spectrum, the greatest risk is associated with release into natural aquatic ecosystems. In between is the placing of transgenic fish into large, commercial outdoor tanks or raceways. This is more dangerous both because it is more probable that the transgenic animals will get into natural water and also because commercial operations are primarily oriented toward profit, not containment. Relevant to the risks associated with intermediate cases of holding facilities would be the presence or absence of physical or chemical barriers to prevent the escape of fish via incoming or outgoing water, possibility of flooding, site security, etc.

Introduction of sterile as opposed to reproductively fertile fish into a natural environment is clearly preferable to introducing animals with reproductive potential. Nonetheless, large numbers of sterile fish can also yield ecosystemic deleterious changes, perhaps over a long period of time, which would be very difficult to reverse.

The authors note that while there is indeed a good deal of data about various cases of nonnative fish being introduced into ecosystems, and transgenic fish are certainly a subset of nonnative introduced fish, there is no general model for predicting what effects a given introduction may produce. Since transgenic fish will differ in significant phenotypic characteristics from their parent stock, they are likely to have a major impact on the ecosystem into which they are placed. Furthermore, the impact will vary in unpredictable ways with the nature of the environment into which the animals are placed and the stability of the ecological community in that environment.

In the face of these considerations, the authors offer some suggestions for research and policy. First of all, transgenic animal lives should be carefully studied until one knows as much as possible about what phenotypic implications attend the genetic changes introduced. Second, research into performance of the transgenic animals should proceed slowly, from closed laboratory situations, to controlled simulations of the natural environment, before any introduction into natural systems occurs. Such model systems should be researched and developed. Improved methods of sterilization should be developed, to preclude accidental proliferation of the engineered gene. Thorough models of risk assessment should be developed after the above concerns are met.

Since this has not yet been accomplished, the authors take strong policy positions, even on aquaculture.

> Pending completion of (1) a thorough program of risk assessment, and (2) demonstration of minimal environmental risk, no introductions of transgenic fishes into production-scale aquaculture facilities, whether public or private, should be permitted. Development and demonstration of a means of guaranteeing sterility of every transgenic fish bound for such facilities may be required, depending upon the results of risk assessments. A federal-level review board of technical experts of various backgrounds should: (1) evaluate research results, (2) recommend decisions on permit applications for use of transgenic fishes in aquaculture facilities, and (3) periodically review and modify the regulatory regime as considered appropriate in light of new knowledge and experience.[38]

Even more adamant is the authors' stand on release into natural waters:

> Intentional releases of transgenic fishes with greatly altered performance characteristics are likely to have considerable impacts upon aquatic ecosystems. Risks of ecological damage may be greater for intentional releases because stocking programs would involve much greater numbers of released individuals. Only if a body of research strongly indicates the safety of stocking a certain type of transgenic fish should proposals for such a program be seriously examined. The burden of

proof should be placed upon those advocating stocking fertile or even sterile transgenic fish into natural systems because, once released, such fish are essentially irretrievable.[39]

In their paper on the regulatory aspects of transgenic fish, the authors orchestrate some of the themes I have already mentioned: the lack of a coherent or systematic regulatory framework covering all transgenic organisms; the need for regulation of commercial as well as academic work on and release of genetically engineered organisms; the domination of decision making by genetic engineers with vested interests and narrow perspectives (although the authors want different sorts of experts more meaningfully involved, e.g., ecologists – they do not mention the public); the need for international coordination of regulations; the need for a unified national law rather than a patchwork of state laws; the need to proceed on a case by case basis in regulation.

The conclusions reached in both the Ecology Society document and the *Fisheries* papers by and large mesh fairly well with my proposal for federal law based on general principles regarding transgenic organisms, arrived at democratically and then democratically implemented on a local level on a case by case basis by committees. The spirit of my proposal differs most radically from the spirit of theirs in the emphasis I place on community input versus their emphasis on broadened groups of experts. I see this difference as representing largely a family quarrel, not a major war. In principle, we are after the same thing: safe and ethical regulation of genetic engineering. In addition, I have a totally different fundamental moral concern that they do not address and I feel is presuppositional to morally sound use of biotechnology, which I shall address in the next chapter.

There is one additional practical environmental concern worthy of mention in association with genetic engineering of animals. By now, we are all familiar with the threat to global and regional ecosystems posed by agricultural expansion in Third World countries. Slash-and-burn techniques deployed in order to provide grazing land for cattle have led to desertification in some areas (Africa) and dramatic loss of species in others (South America). What effect would genetic engineering have on these pernicious pursuits? The answer is not clear. It could be argued that our ability to genetically adapt animals to indigenous conditions would halt such practices while allowing for economic growth. Indeed, thinkers like George Seidel argue strongly that biotechnology can help save the planet, and thus it should be left "underrestricted" rather than "overrestricted." It is equally plausible to suggest that such technology could augment environmental plunder by foisting animals on all sorts of hitherto undisturbed areas with unimaginable consequences, or upset delicate ecological balances. Once again, it is difficult to foresee such risks. But clearly, such considerations should enter into the case by case deliberation discussed above.

POSSIBLE DANGERS: MILITARY APPLICATIONS

A seventh general category of risks associated with genetic engineering is one that we can probably do little about – this is the potential military application of transgenic technology. Since all military weapons development is essentially independent of public scrutiny, this technology also will probably proceed with a life of its own. As much as we might like to believe that nothing could be much worse than some of the biological weapons already in our arsenal – anthrax and plague as well as gaseous "cocktails" of naturally occurring venoms whose molecular complexity makes finding antidotes virtually impossible – genetic engineering could give us worse. Imagine, for example, genetically engineering mosquitoes to spread the AIDS virus, and you can get a sense of this concern. One can imagine all sorts of weapons that could be created using animals as vectors for infecting enemy populations with human pathogens. Nothing short of a universally agreed upon treaty to simply ban transgenic technology in warfare can manage this risk. Even then, one is assured of nothing. Biotechnology is not nuclear technology – it does not take an enormous amount of capital to do transgenic work.

Thus virtually any country – or even any reasonably well-financed terrorist group – could generate biotechnology-based weaponry, provided it could hire the relevant expertise.

This latter point, incidentally, demonstrates the folly of those who, for reasons discussed in the first chapter, would "ban" biotechnology. Even if biotechnology were to be prohibited in the United States, it would not stop – it would simply move somewhere else devoid of restrictions, in the manner of Liberian ship-registration, Third World drug testing, and Caribbean banks for drug money. What this would amount to, then, would be society playing ostrich – genetic engineering would not go away. It would simply be out of sight, and out of oversight. Thus any attempt to deny genetic engineering is to lose the chance to keep it under social control and scrutiny, not to eliminate it. That is indeed why I am writing this book – as a society we can no longer afford to engage spurious, sensationalistic issues of the sort I discussed in chapter 1, while the real issues are essentially socially ignored.

POSSIBLE RISKS: SOCIOECONOMIC CONCERNS

The last set of risks that I shall discuss concerns the socioeconomic consequences of genetic engineering of animals. While these are perhaps more difficult to anticipate than any other sort of risk I've discussed – our grasp of social science is far more tenuous than even our grasp of ecology – we should nonetheless attempt to look at more and less plausible scenarios. For example, it is reasonable to assume, as many small farmers have done, that genetic engineering of animals and other animal aspects of biotechnology, such as widespread use of BST (bovine somatotropin, discussed earlier), would accelerate the tendency for small farmers to go out of business because they cannot compete with large corporations. Small dairy farmers in Wisconsin have argued that, if they needed BST to compete, they could not afford the drug and would thus go under. Similarly, if farmers were forced to acquire patented genetically engineered cows or beef cattle in order to compete, they would be unable to generate sufficient capital to do so, and the domination of agriculture by large companies would be virtually complete.

There are, of course, those who argue that if small farmers cannot compete, they deserve to go under, for small family farms are, in the words of one agricultural economist I know, "archaic, inefficient, outmoded production units, like the mom-and-pop grocery compared to the supermarket." The standard story is that corporatization of agriculture is more efficient, thereby keeping food prices down. This is probably true, though what effect monopolistic corporate domination of the food supply might have on future food prices ought perhaps to be of significant concern.

But there are deeper points to be made here. Even if corporatization of agriculture does keep food prices lower, are we as a society willing to affirm that lower prices are all we care about? To take a telling example, over the past thirty years, universities have moved from being led by senior faculty, exemplars of the ideal of teacher-scholar, to a managerial model wherein control is in the hands of professional administrators, increasing numbers of whom have degrees in "higher education administration," whatever that nonfield may be. Furthermore, they in turn look more and more to business administration and management for the tools to run the university. This in turn leads to bean counting and thinking in business terms about that which in the final analysis is not a business. The result of this emphasis is a loss of quality – classes of three hundred, video classes, and classes taught by graduate students are more efficient, all give "more bang for the buck," to use a popular piece of administrative cant, but also provide poorer education. No wonder, therefore, that our students emerge less and less equipped with the ability to speak, think, and write. The point is that efficiency can be overdone and overstressed at the expense of more fundamental values.

No one would be impressed by a physician or college president who bragged about how many patients or students he "processed"; by an artist whose sole claim to fame was output, or by a person who got rich on child labor. By the same token, there are other values to consider regarding animal agriculture. One is animal husbandry; no one can deny that the price of chicken has stayed essentially the same for twenty years at the expense of the chicken's well-being. Another is holding on to the fundamental American values of honesty, independence, and self-sufficiency that are nurtured and preserved on family farms and ranches across this country. There is ample evidence that Americans – even urban Americans – see the family farm as embodying values worth preserving, not simply as "outmoded economic units." Even as we must see the wilderness as more than just an economic resource, we must see agriculture as more than food factories.

At any rate, these sorts of considerations must be discussed in assessing the risks of biotechnology. What sorts of effects will genetic engineering of animals have on farmers, on workers, on consumers, on urbanization, and, finally, as I shall discuss in the next chapter, on the animals themselves?

In sum, then, I have attempted to discuss some of the general ethical, prudential, and conceptual issues associated with risk, risk assessment, and risk management as applied to biotechnology. I have also tried to delineate a plausible process of democratic deliberation and regulation of genetic engineering to help assure that benefits of genetic engineering will be reaped, that the risks will be minimized, and that the public will be part of the development of biotechnology, not an uncertain, skeptical body to which genetic engineering happens. Finally, I have attempted to delineate some major areas of concern on which public debate and regulation should focus.

CHAPTER 3

The plight of the creature

ANIMAL WELFARE AS A PURELY MORAL CHALLENGE

The sorts of questions considered in the last chapter represent a mixture of ethical and prudential concerns. Ultimately, the weighing of risks and benefits, the anticipation of dangers, and the determination of mechanisms for minimizing and controlling them are dictated by rational self-interest for all parties to the discussion, even those who stand to gain the most from the genetic engineering of animals. Most of the risks I outlined are as much of a danger to the genetic engineer or to the advocate of genetic engineering as they are to the opponents of genetic engineering. Guarding against changes in pathogenicity of organisms, containment of dangerous animals, minimization of environmental despoliation, restriction of monstrous weaponry are all things that any rationally self-interested person would set as a fundamental priority because athey can affect us or those we care about. Indeed, those working in the field of genetic engineering are probably at greater risk than the general public, because of their increased contact with the animals.

Those with a vested interest in the science and technology of genetic engineering are also vulnerable at another level, as has already been pointed out. Any catastrophic outcomes of genetic engineering are likely to eventuate in the imposition of severe restrictions on both research and its practical applications and in wholesale public rejection of biotechnology, as they will play directly into the hands of the doomsayer. Indeed, I have already remarked that a grasp of this point was a major stimulus for voluntary self-regulation by the scientific community during the early days of genetic engineering.

I am not of course suggesting that the ethical dimension of risk management is trivial. Rather, I am pointing out that prudential self-interest will carry along many of those who might tend to ignore strictly ethical considerations when it comes to risk management of genetic engineered animals and thereby achieve the same results via a different route.

Only when we consider the third and final aspect of "the Frankenstein thing" do we in fact encounter a tissue of considerations that require purely moral deliberation, decision, and action, because morality and self-interest are unlikely to coincide in these areas; indeed, as we shall see, they are quite often at loggerheads. In other words, in the area we shall be exploring, one is unlikely to do the right thing for reasons of prudence and self-interest, and, in actual fact, moral behavior in this area will undoubtedly exact costs in self-interest. Thus these issues are the most difficult ones occasioned by genetic engineering of animals, and their resolution correlatively the most vexatious. It is to these questions I now turn.

This final aspect of the Frankenstein myth is more difficult to find than the others in many of the popular renditions of the myth, but it was in fact a central theme in Mary Shelley's novel. This dimension concerns the plight of the creature created through the abuse of science. In the novel, the creature is innocent, yet isolated – shunned, mocked, abused, and persecuted as a result of a state for which it bears no responsibility or guilt. Seeking love and companionship, it finds only hatred and rejection. It has no place, no home, no peers; its life is suffering. One can find traces of this concern for the monster in the original, classic Boris Karloff *Frankenstein* movie, and it is in fact a central theme in the remake of *King Kong* (so much so, in fact, that, when I saw it, the entire movie theater audience was cheering for the gorilla in the climactic final battle with the authorities!).* Translated into our arena of genetic

---

*It is patent in the 1994 Branagh film version of *Frankenstein*.

engineering of animals, this aspect of the myth in essence raises the question of the moral status of animals, of the rights of these animals, of the pain and suffering many of them will inevitably undergo, and of the place of animals in general in our social ethic.

THE TRADITIONAL SOCIAL ETHIC FOR ANIMALS: ANTICRUELTY

The last three decades have witnessed a major change in the Western social ethic regarding the treatment of animals. This change must be understood before one can even begin to address this third aspect of "the Frankenstein thing," for the genetic engineering of animals will be played on a playing field whose rules will ever-increasingly be defined by this emerging ethic for animals.

Though animals were and are property in the eyes of the law, it is certainly true that society has always had some ethic for the treatment of animals, though that ethic has tended historically to be minimalistic. One can find in the Bible, in ancient Greek writings, and throughout the history of civilization, prohibitions against deliberate cruelty toward animals. For example, the Old Testament prohibition against muzzling an ox when it is threshing grain is based on the realization that such an act would involve unnecessary suffering for the animal with no gain for humans. Similarly, the biblical proscription of yoking an ox and an ass together derives from the same insight; it would involve unnecessary suffering for the weaker animal, to be forced to keep up with the stronger, and, again, nothing is gained.

The ethics of a society are, by and large, mirrored in its laws. If one were a methane-breathing Martian anthropology graduate student, sent to earth with a four-hour methane supply with instructions from one's major professor to do a brief survey of U.S. social ethics, one could do much worse than to look at the laws. Whether one is looking at zoning ordinances that restrict sex shops or disallow bars near schools, or at laws against murder, rape, and theft, these laws bespeak our beliefs about right and wrong, good and bad. Beginning early in the nineteenth century, the ancient ethic against cruelty to animals was encoded in similar laws throughout the Western world. These laws were directed against overt, intentional, willful, malicious, deliberate infliction of suffering, and against patent useless neglect. Paradigmatic examples of the former are torturing an animal for fun, tying a firecracker to a cat's tail, burning an animal, or blowing up a frog with an air pump. Paradigmatic examples of neglect are not giving food and water to livestock, not providing shelter to pets in subzero weather, and so forth.

The purpose of these laws was not only to protect the animals from such senseless abuse but, interestingly enough, also to protect humans from people who would do that sort of thing. It has been believed since antiquity, incorporated into Catholic dogma by Thomas Aquinas in the Middle Ages, and reaffirmed by contemporary psychology and psychiatry, that people who begin by abusing animals will often graduate to abusing people. Among the most notorious mass murderers we have seen in the last decade or so, almost all have had histories of animal abuse.[1]

If one examines these laws against cruelty, their judicial interpretations, and their enforcement, one finds that they indeed bespeak a minimal ethic for animals. Their scope is narrow; they are restricted by and large to unnecessary suffering and intentional infliction thereof. What that means, in essence, is that "normal" practices involving animals, that is, practices that are normally done for "human necessity," which really means "human benefit," are exempt from the purview of these laws. Thus "normal" practices involving animals in agriculture, animal research, safety testing, rodeo, hunting, trapping, product extraction, horse racing, circuses, and so forth, are exempt from any constraint by the anticruelty laws.

The reason for the limited scope of these laws lies in the fact that the primary use of animals by humans has always been agricultural, and agriculture traditionally required that farmers utilize good husbandry – keeping animals in environments that suited their natures and for which they had evolved by natural or artifical selection, and augmenting the animals' abilities to thrive by providing food, water, shelter, and so on. In other words, agriculture was an implicit contract between human and animal – "I take care of the animals, and they take care of me," as many traditional ranchers put it. The principle of treating farm animals well and of keeping them in environments to which they were suited was both an ethical and prudential maxim. To harm the animal was to harm one-self, except for brief infliction of pain perceived as absolutely necessary, such as branding of cattle. Thus the social ethic saw no need to focus on "normal" animal treatment and restricted its attention to unnecessary, useless "cruelty."

For that matter, judges have ruled that even such an activity as a tame pigeon shoot, put on by a civic group as a charity fund-raiser, is not covered by the laws.[2] (A tame pigeon shoot is an event in which tame pigeons are released and partici-pants are sold the opportunity to shoot them. Whoever has shot the greatest number when the event is over wins a prize.) In that particular case, the judge ruled that the people who were sponsoring the event were not sadists and psychopaths, were not likely to move on to shooting people, and were putting on the shoot for a good cause, so the cruelty laws were not relevant, despite the manifest pain and suffering experi-enced by the animals.

Not only are these laws extremely restricted in scope and interpretation, they are also given low priority by police and prosecutors, and, if a person is convicted, he or she rarely receives more than a slap on the wrist. Some five years ago, I experienced a dramatic example of the legal system's cavalier attitude toward these laws and toward infractions actionable under their purview. Two of my second-year veterinary stu-dents had acquired a kitten, the possession of which violated their apartment lease. The landlord somehow found out about it, let himself into the apartment with a passkey, beat the kitten to death with a hammer, and left the body in a dumpster. He also left a note for the students, explaining that he had killed the kitten and that they were not allowed to have animals. Understandably upset, the students brought charges against the landlord for cruelty to animals. Some months later, he was tried and convicted – and fined twenty-five dollars. As he left the courtroom, he leaned over to the students and said, with a grin, "For twenty-five dollars, I'd do it again."

Such a situation is not exceptional – any humane society officer can recount similar incidents. These laws were traditionally given low priority; this is a fortiori the case in an already overcrowded legal system, where plea bargaining on major felonies is the norm to keep things moving through the system.

The conceptual limitations of these laws were again dramatically illustrated when the Animal Legal Defense Fund, a group of attorneys whose raison d'être is raising the moral status of animals in society by use of the legal system, attempted to extend the scope of the anticruelty laws by a test case. As animal advocates, they generate many fascinating lawsuits that test, press, and expose the limits of the legal system's control over the treatment of animals. In 1985, they brought suit against the New York State Department of Environmental Conservation, that branch of New York State government charged with administering the use of public lands.[3] Specifically, they charged the department with violating the anticruelty laws by permitting trapping utilizing the steel-jawed trap on public lands. Since there are no laws regulating how often a trapper must check his trapline, an injured animal could be trapped without food, water, medical care, or euthanasia for long periods of time, which, according to the plaintiffs, constituted unnecessary cruelty. They were thus seeking an end to such trapping.

Given the laws, the judge made a wise decision. He opined that the steel-jawed trap was, in his view, an unacceptable device. But given the way the anticruelty laws have been written and interpreted, the actions of the agency in question did not constitute cruelty. After all, steel-jawed trapping is widely done as a means of achieving pest control, supplying fur, and providing a recreational pastime. Thus the activity of trapping is a legitimate one from a legal point of view and does not fit the intent, judicial history, or statutory language of the anti-cruelty laws. If one wishes to change the status of the steel-jawed trap, he asserted, one should therefore go not to the judiciary, but to the legislature. In other words, one must change the laws, that is, the social ethic.

HOW SOCIAL ETHICS CHANGES

Given, then, that the social stage was set for the emergence of a new ethic for animals, how does such an ethic emerge? Here we must recall one of the fundamental dicta of Socrates, namely, that those who engage in philosophical activity about ethics cannot teach, they can only remind. In other words, one does not impart new ethical knowledge to people the way one might teach them the state capitals or the history of music. Rather, one can only get people to reflect on their own ethical beliefs, and on implications of those beliefs they may have failed to note, or inconsistencies among those beliefs that rationality would dictate need to be resolved. This is true whether one is working with the ethics of society in general – for example, on the issue of segregation – or in that area of behavior that the social ethic leaves to the discretion of the individual for example, sexual behavior.

In my own combat metaphor, I have pointed out that when one attempts to effect ethical change in others, one does not employ brute force against the party one is attempting to persuade. (Many of us attempt to bully our children into accepting our ethical precepts and, in fact, only succeed in getting them to reject even more adamantly what we are pressing for.) In other words, ethical argument must be quite unlike the combat represented by sumo wrestling, where each party attempts to push the other one out of the circle by brute force. The only thing that accomplishes is to make one's opponent push back. Thus, when animal rights advocates call researchers "Nazis" and "torturers," this is unlikely to get the scientist to reexamine his or her activities. Similarly, when researchers call their critics "terrorists" and "haters of humanity," this is in turn quite unlikely to generate reflection in those critics.

Thus one cannot put people up against the wall and expect to effect ethical reflection and change. One must rather use judo – utilize the opponents' force against them. In moral terms, this means extracting your conclusions from their premises, showing them that what you wish them to believe is already implicit in their own assumptions.

As one who spends a good deal of time attempting to explicate the new ethic for animals to people whose initial impulse is to reject it, I can attest to the futility of ethical sumo and the efficacy of moral judo. One excellent example leaps immediately to mind. Five years ago, I was asked to speak at the Colorado State University Rodeo Club about the new ethic in relation to rodeo. When I entered the room, I found some two dozen cowboys seated as far back as possible, cowboy hats over their eyes, booted feet up, arms folded defiantly, arrogantly smirking at me. With the quick-wittedness for which I am known, I immediately sized up the situation as a hostile one.

"Why am I here?" I began by asking. No response. I repeated the question. "Seriously, why am I here – you ought to know, you invited me!" One brave soul ventured, "You're here to tell us what is wrong with rodeo."

"Would you listen?" said I. "Hell, no!" they chorused. "Well, in that case I would be stupid to try, and I'm not stupid."

A long silence followed. Finally someone suggested, "Are you here to help us think about rodeo?" "Is that what you want?" I asked. "Yes," they said. "Okay," I replied, "I can do that."

For the next hour, without mentioning rodeo, I discussed many aspects of ethics: the nature of social morality and individual morality, the relationship between law and ethics, the need for an ethic for how we treat animals. I queried them as to their position on the latter question. After some dialogue, they all agreed that, as a minimal ethical principle, one should not hurt animals for trivial reasons. "Okay," I said, "In the face of our discussion, take a fifteen-minute break, go out in the hall, talk among yourselves, and come back and tell me what *you guys* think is wrong with rodeo from the point of view of animal ethics."

Fifteen minutes later they came back. All took seats in the front, not the back. One man, the president of the club, stood nervously in front of the room, hat in hand. "Well," I said, not knowing what to expect, nor what the change in attitude betokened, "What did you guys agree is wrong with rodeo?" The president looked at me and quietly spoke: "Everything." "Beg your pardon?" I said. "Everything," he repeated. "When we started to think about it, we realized that what we do violates our own ethic about animals." "Okay," I said, "I've done my job. I can go." "Please don't go," he said. "We want to think this through. Rodeo means a lot to us. Will you help us think through how we can hold on to rodeo and yet not violate our ethic?" To me, that incident represents an archetypal example of successful ethical dialogue, using recollection, and judo not sumo!

This example has been drawn from an instance that involved people's personal ethics; the social ethic (and the law that mirrors it) has essentially hitherto ignored rodeo. But it is crucial to understand that the logic governing the above case is precisely the same logic that governs changes in the social ethic as well. Here also, as Plato was aware, lasting change occurs by drawing out unnoticed implications of universally accepted ethical assumptions.

An excellent example of this point is provided by the civil rights movement in general, and, more particularly, by Lyndon Johnson's shepherding of the monumental Civil Rights Act of the 1960s. As an astute politician, and particularly as an astute Southern politician, Johnson had his finger on the pulse of how American segregationists were thinking. He realized that the social zeitgeist had progressed to the point that most Americans, even most Southerners, accepted two fundamental premises, one ethical and one factual. The ethical principle was that all humans should be treated equally in society, and the factual assumption was that blacks were humans. The problem was that many people had never bothered to put the two premises together and draw the inevitable conclusion, namely, that blacks should be treated equally. Johnson believed that if this simple deduction were put into law at this particular point in time, most people would "remember" and be prepared to bow to the inevitable conclusion. Had he been wrong, the Civil Rights Act would have been as meaningless as Prohibition!

We have, in fact, over the last forty years, lived through a good deal of Platonic ethical recollection regarding the ignored consequences of our accepted social ethic. We have seen that ethic rightfully extended not only to blacks, but to women and other disenfranchised minorities, when there was no morally relevant basis for withholding that ethic. To deny an otherwise qualified woman admission into veterinary school, for example, on the grounds that she is a woman (a practice that was rife in these schools until the late 1970s), is as much a violation of the implications of our social ethic as was segregation. Nonetheless, getting people to recollect is a long, hard process, despite the simplicity of the argument on paper. But, still and all, it has occurred, and society has been very much sensitized to recollection regarding those groups of people hitherto disenfranchised and ignored.

PROBLEMS IN THE ANTICRUELTY ETHIC

What is essentially happening in society regarding animals, then, is a growing dissatisfaction with the venerable ethic that underlies the laws I have been discussing, the ethic of kindness and cruelty. These two words tend to exhaust our limited social vocabulary for discussing the treatment of animals. Even the humane societies and animal welfare organizations, those groups specifically chartered to protect animals in society, have tended to be guided in their discussion of these issues by the dominant notions of kindness and cruelty. Thus one sees mailings from animal welfare groups that demand, for example, that the testing of cosmetics on animals be stopped, and this point is almost always articulated as "Stop the cruelty!" The use of animals in biomedical research is characterized as "cruel"; so described also are any abusive uses of animals, be they dog and cock fighting or rodeos, circuses, hunting, confinement agriculture, or whatever. Correlatively, those who treat animals properly are said to be "kind" and elementary school children are peppered with coloring books, signs, and posters that enjoin "kindness to animals."

Why is this emphasis on cruelty and kindness misguided? In the first place, as we shall shortly see, most animal abuse does not arise out of cruelty, and in fact many people – for example, ignorant pet owners – have the kindest intentions but nonetheless cause harm to their animals by not understanding their needs and natures. Second, most people who cause animal suffering are usually motivated by reasonable motives – curing disease, advancing knowledge, producing cheap food, and so on, not by cruelty. And describing them as cruel (or malicious or sadistic) generally does not help to generate critical self-examination. What it in fact does is cause defenses and rationalizations to spring solidly into place and battle lines to be drawn. Third, the issue of how we should morally behave toward animals is a question of moral obligation – what we ought to do or are obligated to do – not a matter of "kindness." Kindness suggests an overflowing of good will, above and beyond the call of duty. Yet questions of obligation are questions of duty. Kindness essentially becomes a patronizing notion if it is the only reason for treating something morally – imagine someone, in the civil rights or women's movements, choosing as a battle cry "Be kind to women" or "Be kind to blacks"! How many women would be satisfied with being admitted to medical school solely on the grounds that the admissions committee has decided to show kindness toward women?

Kindness and cruelty are basically inadequate for handling the moral issues surrounding the proper treatment of animals! To support this claim, I ask the reader to do a little thought experiment. Consider a pie chart that represents the total amount of suffering and pain animals experience at human hands. And ask yourself what percentage of that suffering is the result of intentional, sadistic, malicious, willful cruelty of the sort covered by the anticruelty laws? The answer, of course, is that, on the chart, that sort of activity would not even be represented by a slice, but only by a very thin line. In other words, most animal suffering is not the result of people trying to inflict suffering, but rather the by-product of people pursuing socially acceptable, even desirable, goals – producing cheap food derived from animals, advancing knowledge and curing disease (human and animal) through animal research, testing the safety of products intended for humans, and so on.

This realization is the key to understanding the new ethic for animals. Society is demanding the limitation and control of animal suffering, regardless of its source – benevolent or malevolent. The traditional anticruelty laws, and the ethic they mirror, are clearly inadequate for handling this task, as we saw graphically illustrated in the steel-jawed trap case discussed above. So society is groping for a new ethic for animals in an inchoate way and, as I shall demonstrate, the direction in which it is moving is both rational and inevitable.

THE GROWING SOCIAL CONCERN WITH ANIMAL TREATMENT

It is quite evident that concern for animals, their well-being, and their suffering does represent a major social-ethical concern, not only in the United States, but throughout the Western world. According to a colleague of mine at the National Institutes of Health, in the late 1980s Congress got more letters, telegrams, and phone calls on animal welfare–related issues than on anything else. The National Cattlemen's Association reports a similar state of affairs. Unprecedented levels of social concern and pressure forced the passage in 1985 of two pieces of legislation designed to assure the well-being of laboratory animals, where previously researchers had essentially enjoyed carte blanche in animal use.[4] When the Department of Defense announced in 1983 that it was going to shoot a group of anesthetized dogs in order to teach trauma surgeons how to manage the wounds inflicted by tumbling bullets, the department received more irate letters, cards, and phone calls on this matter than on anything else in its history. Zoos and aquariums have been forced by public pressure to create environments more friendly to the animals. Animal agriculture sees animal welfare as one of the three major problems facing the industry in the next century. Congressman Charles Stenholm, a prominent member of the House Agriculture Subcommittee, has predicted that there could well be federal legislation for assuring the well-being of farm animals by the year 2000.[5] In 1985, Britain adopted a sweeping new law governing animal research, the first major law since the pioneering act of 1876. Other countries – Holland, Australia, and South Africa, for example – have adopted or committed to major new laws governing animal experimentation. New laws severely limiting confinement agriculture have been adopted all over Europe and are pending in the EC. The Canadian baby harp seal slaughter was stopped by public outrage that led to a European boycott of Canadian fur products. Major cosmetics companies such as Avon have announced that they no longer test their products on animals; other companies are spending millions of dollars in the search for alternatives. Tuna companies have ceased to use nets that entangle and kill dolphins. Eight out of ten readers polled by *Parents* magazine affirmed their belief that animals have rights.[6] Hundreds of pieces of legislation pertaining to animal well-being are introduced in Congress and state legislatures annually.

The above list could be continued almost indefinitely, but the point has been made. Animal welfare is currently and has been, roughly since the middle 1980s, a major and constantly growing social concern – one that appears to have entered the social mind only recently, but one that has nonetheless flourished and spread.

Before examining in more detail the nature of this ethic and its relationship to the genetic engineering of animals, it is worth considering a question that I am frequently asked by people in animal research, agriculture, hunting, and other areas that depend on animal use. The question is: "Why, after hundreds of years during which our treatment of animals was barely a concern and prohibitions against cruelty sufficed to define the social ethic, has this new concern surfaced and pushed social morality forward at an unprecedented rate?" While I can generate no conclusive answer, it is evident that certain factors have played a major role.

First, we as a society have gone through four decades of ethical soul-searching, focusing our moral concern on groups that were traditionally ignored, unfairly treated, and exploited – women, blacks, Hispanics, and other minorities, children, the aged, the Third World, the handicapped, and so on. In other words, the prevailing zeitgeist leads people to look for injustice and try to rectify it, not sweep it under the carpet or appropriate it for one's own benefit. Many of the most active people in the forefront of the animal rights movement are people who were also active in earlier struggles for justice. Correlatively, many people in other social movements – the women's movement, for example – see the callous exploitation of animals as part of the same systematic oppression that they themselves have attempted to combat.

Second, animal issues have been kept prominently in view by extensive media coverage because, as one reporter told me candidly, "animals sell papers." Whether the story concerns the infamous head injury tapes graphically documenting atrocious behavior toward animals by researchers, or is a series of syndicated five-minute pieces on confinement raising of veal calves, or documents the poaching of elephants or whales trapped in an ice floe, such stories have enormous impact and generate vivid images that are not easily forgotten.

Third, our society has become an overwhelmingly urban and suburban one; only about 1.7 percent of the population is now engaged in production agriculture, and even fewer in animal agriculture. Agriculture, once the predominant occupation in society, is now essentially invisible. And even though the largest number of animals consumed in society are agricultural animals, few of us have any direct contact whatsoever with how they are raised, killed, and processed.

Where once virtually every member of society had a relative or friend engaged in some area of agriculture, such is no longer the case, and dramatically no longer the case.

Imagine taking a time machine back one hundred years and visiting any part of the United States, urban, rural, or frontier. Suppose, further, one were to stop people in the street, at random, and ask them to associate the first word that comes into their mind with a word you utter. If you said "animal," they would very likely say "farm," "cow," "horse," "food," or something along these lines. Today, however, if one were to try a similar test, in all but intensely agricultural areas, one would get very different responses. If one said "animal," one would probably get in response "pet," "friend," "companion," "dog," "cat." (Over 90 percent of the pet-owning population say that their pet is a member of the family.[7]) This thought experiment illustrates the degree to which the conceptualization of animals has changed in current society.

In the face of this change, it is no surprise that people are more concerned about animal pain, suffering, abuse, and treatment than ever before. The root image of animals has changed dramatically from the "beast of burden" or "food animal" to the pet. It is bitterly amusing that a major producer of laboratory animals has attempted to wean animal biomedical researchers away from dogs – specifically beagles – by developing and pushing miniature swine as an alternative about which, in essence, no one cares, only to find that the very same animals were being bought – and cared for – as pets.

Fourth, the animal issue has been fueled and sparked by a large number of intelligent and articulate people, primarily philosophers, who have expressed the issues in sophisticated terms far removed from the traditional language of "kindness" and "cruelty." These philosophers – Singer, Regan, Sapontzis, Clark, Jamieson, Pluhar, myself, and many others – have given the public a rational framework for articulating and defending what was traditionally seen as merely a matter of emotion and sentiment – and thus easily dismissed. The burden of proof is now upon those who would exploit animals with no constraints and who would schematize animals as merely tools for human use. In my own work in this area, I have always felt that I am articulating in philosophical terms what is happening in society, not developing a new ethic from whole cloth. I will return to this point.

Fifth, and undoubtedly most significant in fueling public concern about animals, has been the relatively recent change in how animals are used in society, a change that occurred primarily in the last fifty years. Prior to that, as I said, the overwhelming use of animals was in extensive agriculture, wherein the animals lived their lives under conditions for which they had been biologically evolved or bred. Animal agriculture was, to a large extent, animal husbandry, care for the animals, managing the animals under conditions that fit the animals' natures. As livestock handling specialist Temple Grandin has said, agriculture was a fair contract, with humans helping animals to live their natural lives (by providing shelter, food, water, protection from predators, etc.), and animals providing their products or their lives. Anyone who abused their animals, or tried to raise them under conditions for which the animals were not biologically fit, would rapidly go out of business – the animals would die, or not produce, or not gain weight, or get sick. Thus it is fair to say that the sort of activity (indeed the only activity) that required moral and legal attention was indeed cruelty and malicious abuse; because the systematic use of animals of necessity had to accord with the animals' physical and psychological natures. (One exception to this was animal research, which, in quantitative terms, was performed on very small numbers of animals. Nonetheless, as mentioned earlier, it was being regulated in Britain by 1876.)

In the last fifty years, however, things have changed dramatically. Animal research and testing, for example, has increased significantly as an industry and often involves deliberate infliction of pain or disease on animals. Even more revolutionary, though, was the change in the nature of animal agriculture, which became industrialized, with industrial methods and high technology replacing extensive management and husbandry. Symbolically, university departments of "Animal Husbandry" changed their names to departments of "Animal Science." Intensive or confinement agriculture – "factory farming" – was born. In this kind of agriculture, technology and science allow us, as it were, to put square pegs in round holes, to keep animals under conditions for which they are not biologically or psychologically suited, and still make a profit. Whereas any farmer foolish enough 100 years ago to try to raise laying hens in cages would have been out of business in months or weeks because of brushfire-like disease spread, today's farmer can override this natural check on confinement by use of antibiotics and vaccines. However unhappy the animals may be, they can still be productive. Thus the traditional contract between human and agricultural animals – we help you live a decent life, you give us your products or life – was broken. We can now fail to respect the animals' natures and still be productive and efficient. And with that failure has come significant amounts of new suffering, such as depriving animals built to move of the ability to move.

It is fair to say that revelations about this massive change from husbandry of farm animals to factory production of them, which began seriously after World War II, is the single factor most responsible for the social demand for a new ethic. When, during the 1960s, the British public – and the world – was informed about what had happened to animal agriculture, largely through the publication of Ruth Harrison's pioneering book *Animal Machines*,[8] the response was unprecedented. Here was massive animal suffering on an unimagined scale that was the result of a quest for efficiency and productivity, not a result of cruelty. And so new ways of thinking about animal abuse and our obligations to animals were called for. I shall discuss this point and this case further when I develop the ethic that has since emerged. Here it is enough to note that changes in animal use and production were highly significant for creating the demand for a new ethic for animals.

BEYOND CRUELTY: THE NEW SOCIAL ETHIC
FOR ANIMALS

In any case, I can now draw the diverse pieces of my discussion together so that the reader can see the inevitability of the ethic that has emerged for animals. I have argued that the traditional ethic proscribing cruelty is significantly inadequate in current society, because people have come to understand that the overwhelming majority of animal suffering today is not the result of deliberate cruelty, but is rather a by-product of activities aimed at producing profits, advancing knowledge, improving health, supplying cheap food, and so forth. For most of history, the major use of animals in society was agricultural, and traditional agriculture depended on fitting the animals and their living conditions to each other, and on husbandry. Thus, in the past, most animal abuse statistically may well have been cruelty, and the traditional ethic (and laws) sufficed. Today, however, burgeoning science and technology have made agriculture, which still represents the major use of animals in society, less contractual between human and animals and more exploitative, less a matter of husbandry and more analogous to industrial production. (Indeed, animal science prides itself on applying industrial methods to animal agriculture.) The result is what is popularly known as factory farming, which does not allow the animals to live the lives for which they are biologically suited. This has ignited a demand both for reform and for a new ethic all over the world.

In addition, research and testing has emerged in the last half century as a major and highly controversial area of animal use, essentially also calling for a new ethic, since researchers are not sadistic or cruel people, but again cause suffering. As society looks for a new language beyond talk of cruelty and kindness to express its moral concern for animals, it is natural and inevitable for it to look at what it already believes to be right and wrong, and at other parts of the ethical machinery it already uses in nonanimal contexts. As it happens and as I have stressed, we are emerging from forty or so years of extending our social ethic to disenfranchised humans, to whom our mainstream ethic for "first-line" humans was not fully applied for hundreds of years. In such a context, it is inevitable that society would look to our ethic for humans (mutatis mutandis, as philosophers say – that is, appropriately modified) to accommodate our growing moral concerns about the treatment of animals!

Am I saying that society is moving beyond cruelty by equating animals with people? No! What I am stressing is that society is taking some of the ethical machinery we use to think about the treatment of people and is, as it were, exporting it to the domain of animal treatment. And I further believe that society, common sense, and those philosophers like myself who try to articulate what is and ought to be occurring in society, have homed in on a very appropriate portion of our ethical machinery to use for dealing with animal issues, as I shall now explain.

Our society, like all other societies, is faced in human ethics with a problem that cannot be avoided, growing out of the inevitable conflict that arises between respect for individuals and respect for the general welfare or general good. Different societies have solved this problem in different ways. Totalitarian societies, like Hitler's Germany, come solidly down on the side of what they believe to be the general good, with no respect for individuals. At the other extreme, some of the hippie communes of the 1960s put total emphasis on respect for individual desires, with the inevitable result, of course, being the compromising of the group's welfare and even the group's existence.

The best answer, I believe, has been devised by our society. While we indeed make most of our social decisions by weighing what is most conducive to promoting the general welfare, we build certain fences around individuals to protect them from being oppressed, crushed, and submerged by the general welfare. Thus we do not allow the torture of a bank robber to reveal where he has hidden the money, or even the torture of a terrorist who has planted a bomb in a kindergarten. We do not allow seizure of property without compensation for the public good, censorship of speech, or restriction of belief. In other words, these fences protect even from the general welfare those interests that are believed to be essential to and constitutive of human nature – speech, religion, protection from torture, and so forth. The name we give to these partly legal, partly moral notions is *rights* – rights protect the integrity of human nature, as it is found in individual humans, from being sacrificed for the sake of general expediency.

The emphasis on rights, on assuring through the legal system that fundamental aspects of the natures of humans are protected is a major feature of general social thought over the last half century – witness concern with the rights of minorities, women, children, native peoples, students, the homeless, and so on. One of the key dimensions of this rights thinking is the stressing of our obligations toward traditionally disenfranchised groups and individuals, obligations that are based in duty, not merely benevolence. In my view, rights talk in society has served to stress the morally relevant similarities between those who have traditionally not enjoyed full standing or concern in the moral arena and those who have. There have, for example, always been people who were sincerely motivated by benevolence – or kindness – to allow women into veterinary schools (from which they were mostly excluded until the late 1970s). The emphasis on women's rights, however, stresses that whether or not one feels benevolently disposed toward women, many of them have the same characteristics demanded of men who are admitted to veterinary schools, and thus we are obliged, as a matter of justice, to admit them – they are entitled to admission.

Although some people in society wish to see the notion of rights that we use for humans exported in toto to animals, so that no amount of human benefit justifies harming animals (a full or strong animal rights position), I do not believe that is what mainstream social thought has as yet evolved to. However, the notion of rights does play a significant role in shaping new social-ethical thought about animals. A survey of its mainstream readership by *Parents* magazine revealed that 80 percent of that group believes that animals have rights, though 84 percent also believes that it is permissible to use animals for human benefit.[9] When I talk to Western ranchers, who also clearly believe that it is permissible to use animals for human benefit, over 90 percent of the more than five thousand I have informally polled affirm that animals have rights.

So, clearly, some message important to current social thought is being carried by the notion that animals have rights. If one attempts a rational reconstruction of what is taking place, I believe the notion of rights captures the following features of the emerging ethics for animals we have been discussing:

1. A belief that the proper treatment of animals is a duty, not a matter of choice dependent on whether the agent has a benevolent attitude toward animals; the latter characterized the kindness-cruelty view.
2. As a duty, proper treatment of animals should be legally encoded (i.e., be mandatory).
3. Given the recent (mid–twentieth-century) developments in confinement agriculture, biomedical research, and toxicology, which, unlike traditional agriculture, can and do work against the animals' natures, mandated regulation must take the place of the traditional "contract" that characterized nontechnological animal agriculture to assure respect for animal interests flowing from their natures.
4. All future and new use of animals in society should be regulated so as to assure that humans do not continue to cavalierly encroach on animals' well-being for human benefit. In other words, the effect on animals should be considered; not just the benefit to humans.
5. The well-being that should be protected involves both control of pain and suffering and allowing the animals to live their lives in a way that suits their biological natures. Thus Sweden has passed a law for agricultural animals that mandates that all systems of keeping farm animals must first and foremost accommodate the animals' natures. For example, the law grants cattle "the right to graze" in perpetuity and abolishes the confinement raising of pigs and chick-

ens in which the animals are not permitted to move naturally.[10]

6. Society is morally concerned with the well-being of all animals, rather than just the "favored" dogs, cats, and horses
   that dominated traditional concern for animals, even on
   the part of humane societies.

7. Society demands control of animal suffering even when
   such control costs society more than just a minimum, and
   even when "efficiency" is sacrificed. This is a major step
   beyond the traditional concern for suffering that was the
   result of deviance and sadism and produced no benefit for
   society in general.

One could argue that, in the face of the fact that society
continues to approve of the use of animals for human benefit,
even if it now demands that we protect the animals' fundamental interests as determined by their natures, we are merely augmenting the traditional "animal welfare" or kindness
and cruelty ethic, not changing it in any fundamental way. I
believe that such a view does not capture the major changes
listed above. The notion of enacting a host of new laws to
cover all aspects of animal use, to the benefit of the animals, is
radically different from what occurred in the past.

Ultimately, it is a very different ethic to worry about all
animal suffering, even if that suffering is a by-product of
using animals in normal activities for human benefit (science,
agriculture, product testing), than to worry only about suffering resulting from deliberate sadism or purposeless neglect.
So I would argue that even if, philosophically, society does
not yet have a full or strong rights ethic for animals, the way in
which it is looking at welfare does bespeak the significant
influence of the concept of rights. So too does the demand
that we protect the animals' natures in our use of them –
despite the fact that such protection may impair our efficiency
in using them.

Ultimately, then, we can think of the new social ethic as one
in which human use of animals is checked, modified, and
constrained by the concept of rights. We can think of this

either as a weak version of a rights ethic, or as a welfare ethic strongly influenced by the notion that animals, like humans, have natures, and that respect for the basic interests that flow from those natures should be encoded in our social morality.

As ordinary people know well, animals too have natures, genetically based, physically and psychologically expressed, which determine how they live in their environments. Following Aristotle, I call this the *telos* of an animal, the pigness of the pig, the dogness of the dog – "fish gotta swim, birds gotta fly." Animal telos is of course not the same as human telos; thus the protections they require are not the same, thus the rights of animals cannot be the same as the rights of humans. But the fact that animals do have interests that are as important to them as speech and belief are to us is indubitable. Social animals need to be with others of their kind; animals built to run need to run; these interests are species specific. Others are ubiquitous in all species with brains and nervous systems – the interest in avoiding pain, in food and water, and so forth.

In any event, given that most current animal abuse is not cruelty, but rather arises out of nontraditional ways of exploiting animals for human benefit (or even for the benefit of other animals), the rights portion of our social ethic is most appropriate to guide the new social ethic for animals. Recall that rights protect individuals and their natures from being eroded for the common good. This sort of erosion is precisely what we have seen characterize current animal abuse – exploitation for the general welfare at the expense of basic interests of the animals. The concept of rights provides a model for legal/moral protection from such exploitation and can be appropriately demanded for animals as well.[11] The concept of rights further provides a framework for socially demanding the same respect for animals' natures that was of necessity built into traditional agriculture.

I do not believe that society as a whole is saying "Do not use animals at all," though a minority of people are saying precisely that. But I do believe that society as a whole *is* saying that, if we use animals, they should not suffer and should be happy. If we raise animals for food, they should live happy lives under conditions that meet their biological natures. If we use animals for research, they should not suffer. If we exhibit animals in zoos, they should live as happily as possible under conditions where they can express the powers and activities that make up their natures and their lives. Sadly, there is, in my view, no area of animal use in human society where animals are getting the best treatment they could possibly get, even consonant with that use.

What I have just described is, I believe, the essence of the mainstream, worldwide, emerging social ethic for animals and is the sense in which the concept of *animal rights* has entered general social thought on animals. The issue is abuse and suffering, not cruelty; justice and fairness, not kindness. The vehicle for assuring justice is the law, specifically as guided in human ethics by the concept of rights as protection for a being's nature or telos.

Another way of schematizing this ethic is to say that it recognizes that animals are "ends in themselves," as Kant said of humans, not just means to our ends. What we do to animals matters to them, not just to us. In this fundamental moral respect, animals are like human persons, not like tools. We surely use other people for our ends, but we must always keep their interests in mind as well. It has become a cliché in discussions of sex that immorality in sexual behavior is tied significantly to how we treat the other person. We must recognize that one's sexual partners are not merely there for one's pleasure, but also for their own. This is the difference between a sexual partner and a sexual toy. Just as we cannot ignore the needs of other humans in our interactions with people, so we cannot ignore the needs of animals in our interactions with them.

The emergence of this ethic in society has, as mentioned earlier, prompted action on animal issues all across the Western world in all aspects of animal use. With the growing awareness that animal suffering has reached massive proportions in fields like agriculture, research, and testing, people have demanded checks on that suffering. Even those with vested interests in opposing constraints on animal use have begun to understand the basis for those demands. For example, I recently attended a major international conference on the welfare of food animals, where the full gamut of opinion on the issues was well represented. A person speaking for industry – that is, for those who have a vested interest in confinement agriculture – made what I thought was a startlingly honest statement to the whole group. These confinement agricultural systems, he asserted, were developed after World War II by the industry for the sole purpose of producing inexpensive and plentiful food. At that time, he continued, no one was talking of animals' psychological needs, natures, or rights. So the industry forged ahead and developed high technology confinement agriculture. Now they are confronted with the social demand that they factor in all aspects of the animals' welfare – beyond food and water – into their production systems. Obviously they must do so, he added, but it will take time and research, as these considerations are relatively new.

The research community, also, was not prepared for the mainstream concerns that emerged over the past twenty years about research animals. Over a period of years, they reasoned, unrestricted animal research has produced many health benefits for humans. Almost no one except "extremists" spoke of animal rights for most of the history of animal research, so the current demand must also be a function of loud, headline-grabbing extremists. They of course failed to understand the social dynamics we have discussed leading society in general to a new level of awareness and concern about animal suffering. This lack of vision was further compounded, as we discussed earlier, by scientific ideology, which denied the relevance of moral issues to science.

By now, however, the scientific community should be aware of the nature of the new ethic emerging for animals, as should all animal users, for it has been "writ large" into the legal system. In my view, the new U.S. laws on care and use of laboratory animals, as well as major new laws passed elsewhere in the world, collectively provide an excellent indicator of where social thought stands on animal well-being, and what society will and won't accept in the present and future.

As mentioned earlier, I was part of a group of Colorado citizens – a philosopher, an attorney, and a number of distinguished veterinary scientists – who, in the mid-1970s, began to draft legislation for laboratory animals that went beyond the traditional anticruelty approach and was guided by the sort of ethic outlined above, which we correctly saw was emerging in society. Members of the group collectively enjoyed over sixty years of animal research experience – indeed, one of us, the late Dr. Harry Gorman, had invented the artificial hip joint used in humans and animals and had also led the animal experimentation portion of the space program. Although our views of animal experimentation varied greatly, we all agreed on a core ethic we wished to encode in laws.

In the first place, we felt strongly that if animals were to be used in research, they ought not suffer pain, distress, fear, anxiety, or any sort of misery. Analgesia, sedation, and anesthesia should be mandated, and the traditional ideology of science agnosticism about animal pain and consciousness should be eroded. We also believed that animals should not be used over and over, should be euthanized if they were suffering, and that how they were kept and housed should respect their biological natures. If they were to be euthanized, they should have the best death possible.

We further believed that law should not only enunciate fundamental principles of ethics, it should also aim at educating scientists and breaking the ideological bonds that perpetuated the views that science was value-free and that animals lacked minds and feelings. Toward this end, we advocated "enforced self-regulation" through the vehicle of local animal care and use committees, which reviewed protocols for research and monitored animal care and use. By asking researchers to discuss such issues as control of pain and suffering, justification for animal use, and so on, vis à vis each piece of research to be done, we felt scientists' thinking would gradually evolve to the point where these and other ethical concerns were second nature to them, and moral concern for animals would be internalized. We hoped that, in this way, scientists would plug into the emerging mainstream ethic for animals and that the traditional ideological distancing of science from ethics would be undercut.

Despite vigorous opposition from powerful elements in the research community, who attempted to frighten the public with the claim that law would slow down medical progress, two federal laws based on this model were passed in 1985,[12] antedating my own guess of when this would occur by fifteen years! Clearly, the social ethic was stronger than we had thought. Although not everything we had delineated was passed (for example, rodents used in industry are still not covered by either law), most of what we argued for was legislated, at least partially. (Industry is, in any case, forced by evidence of public concern, if not by the law itself, to be state-of-the-art, even when this is not legally required.) Most important, the law, in essence, acknowledges the right of animals not to suffer physically or mentally in research, unless (as is rarely the case) one is studying pain and suffering, or control of pain and suffering would vitiate the experiment. (This occurs in less than 10 percent of research protocols.) Proper euthanasia is mandated, as are proper use of anesthetics, analgesics, and tranquilizers. Protocol review and oversight by local committee is mandated and, although accommodations suited to all research animals' natures are not currently required, the law does mandate exercise for dogs and living environments for primates that "enhance their psychological well-being."

One could argue that this law is so weak it hardly provides evidence for rights thinking, certainly not on a par with the rights we accord humans. Human rights, after all, are not so easily trumped. But to make such an assertion is to ignore the historical context in which the law arose. In this context, research use of animals was, by legal fiat, immune from the cruelty laws that articulated the social ethic. Nothing one could do to a research animal in a university could count as cruelty. Furthermore, one could find virtually no use of analgesics to control pain in research animals, no scientific literature recommending that use, and an ideological denial of the reality of felt pain in animals.[13] The research community enjoyed total laissez-faire in animal use and much of that community vigorously resisted any change in the status quo, as I saw firsthand when I testified at congressional hearings on behalf of our legislation. The idea that animals are worth caring about in their own right rather than merely as tools for accomplishing a research objective was radical in such a context, and it has made a major difference to the amount of pain that research animals suffer. The idea that research animals are entitled to a living environment that suits their natures is again revolutionary and has led to major changes in primate care. We must recall, too, that the law reflects a compromise between societal concerns and the desire of the politically powerful biomedical research community to be left alone. Society would, I believe, like to be in a position to outlaw certain experiments and types of experiments and would like to see research proceed without inflicting pain and suffering on animals at all.

If one discusses the new laws with researchers, with members of the public, with leading regulators, as I have done, one gets a clear sense that the law is perceived as encoding some rights for research animals – the right to have pain controlled, the right to an enriched environment that suits them, not just us, the right to a good, painless death. And it is significant that researchers and committees are extending these notions to areas where they are not legally mandated.

I conclude, therefore, that all of this betokens a step far enough beyond the traditional kindness-cruelty ethic so as to represent a moral quantum jump. Whereas the traditional ethic for animals forbade "unnecessary suffering," it allowed "necessary suffering," which it defined weakly as suffering that was inconvenient or expensive to alleviate in the context of human use. The new ethic is a good deal stronger – the research laws allow as "necessary suffering" only that suffering that is impossible to alleviate, in the context of human use. Expense or inconvenience or inefficiency is no longer an excuse for not controlling pain and suffering. And this approach has major implications for other areas of animal use. I have already mentioned the Swedish law, which the *New York Times* called an "animal rights law."[14] The other major conceptual breakthrough in these laws, of course, is the acknowledgment that animal suffering that is not the result of overt, sadistic cruelty is also worthy of social moral attention.

In sum, then, although the new ethic for animals is surely not a full-blown rights ethic, it moves significantly beyond our traditional minimalistic social ethic for animals. It is, perhaps, best schematized as a much augmented welfare ethic, informed by the ideal of respecting rights flowing from the animals' natures. Like our ethic for humans, it is thus a mixture of utility and rights thinking, albeit that the latter appears in a much attenuated form. Nonetheless, it is a foot in the door, as it were, and it is not unreasonable to foresee such rights evolving in strength over time, even as those of disenfranchised humans evolved.

Scientific ideology is certainly being eroded by these new laboratory animal laws. If the law says that animal pain must be controlled, scientific ideology cannot continue to maintain that there is no such thing! An amusing anecdote evidencing this point has been related by Dr. Robert Rissler, then the USDA APHIS (Animal and Plant Health Inspection Service) administrator charged with promulgating regulations giving operational meaning to the law. According to Rissler, he was, as a veterinarian, much perplexed by the requirement of "enhancing primates' psychological well-being." Traditionally trained veterinarians do not learn much about primates, and "psychological well-being" does not come up in veterinary education. So Rissler decided that he would approach the American Psychological Association for help – surely they would know. But he was disappointed – the experts at the association told him that there was "no such thing" as "psychological well-being of primates." "Well," he replied, "there will be after January 1, 1987" (the date the law took effect), "whether you people help me or not!"

To the credit of the research community, animal care committees are often extending their concern beyond what is mandated, as I have mentioned. Thus, for example, many committees demand that the control of pain be extended to animals who are not covered by either of the laws – for example, invertebrates and farm animals used in agricultural research. Investigation into the nature of pain, suffering, distress, anxiety, and so forth in animals has been sparked by these laws; indeed, the National Academy of Science has recently issued a book dealing with these issues. I have recently edited a book about the various laboratory animal species where the authors, all well-established scientists, discuss, among other things, environments for the various types of animals that would be oriented toward making them happy and meeting their psychological as well as physical needs.[15]

Although federal legislation in the United States has thus far been promulgated only in the area of laboratory animals, other fields of animal use have also experienced the influence of the new ethic and have undergone changes that reflect the power of that ethic. As mentioned earlier, a number of cosmetics companies have announced that they no longer employ animal testing. The fur industry has been dealt a crippling blow by the ever-increasing view of fur as an immoral use of animals. Some fairs and rodeos held in areas where there are large urban populations and even in rural areas have abandoned events like greased-pig contests, and they have instituted rules against jerking calves in the roping events. Invasive use of animals in science teaching at the elementary and secondary levels has been virtually eliminated. At the college level, invasive laboratories have been minimized in many universities, and those students objecting to them, even in required courses, are ever-increasingly permitted to elect an alternative. Even in medical and veterinary schools, many schools now offer programs that meet student moral commitments to learn medicine and even surgery without hurting or killing animals.

Zoos and aquariums have been under pressure for over a decade to make their exhibits and enclosures "animal-friendly," thereby eliminating the suffering and resultant behavioral anomalies often occasioned in animals by cramped, austere, and boring environments. In Canada, the Ministry of Fisheries and Oceans refused permission for an aquarium to take killer whales from Canadian waters until the aquarium could demonstrate that it had built accommodations for the animals that suited their telos. At a recent conference I attended, participants from agriculture, science, industry, and animal welfare were involved in a workshop on animal ethics. We were charged with reaching some consensus positions. We did so by trying to demarcate examples of animal use that we all agreed were beyond the pale of acceptability. Much to my surprise, we all agreed that training bears to ride bicycles or dance in circuses was unacceptable, an example suggested not by the animal welfare people, but by a man from agribusiness.

Perhaps the most dramatic evidence of the new ethic was provided by the passage of the sweeping new law for agricultural animals in Sweden I mentioned earlier. In the late 1980s, the Swedish Parliament passed an animal welfare law for farm animals that is the strictest in the world yet passed through Parliament "virtually unopposed."[16] The main thrust of this law is virtually a paradigmatic instantiation of the new ethic I have described. What the law mandates is that farm animals be allowed to live their lives in accordance with their natures, or telos as I have called it. Indeed, the entire bill is informed by the notion of rights I discussed. While acknowledging that people will eat animals and animal products, the law reaffirms the ancient idea of husbandry – that cattle have the right to graze; that chickens and pigs have the right to freedom of motion; that animals who would naturally use it have a right to straw; that animals have the right to separate feeding and bedding places; and so forth. Drugs such as antibiotics can only be used to treat disease, not to conceal the untoward effects of confinement. Slaughtering must be as painless as possible.

Clearly, the law is designed to do two things essential to the ethic I have described. First, it guarantees farm animals the right to as pain-free an existence as possible at human hands. Second, it addresses not only overt physical pain, but also the sort of suffering that results from the failure to adjust the way the animal is kept to its biological telos. And Sweden is just a sentinel for a worldwide movement – similar, albeit less dramatic, reforms are being demanded all over Europe, in the EC, and in North America.

GENETIC ENGINEERING AND THE NEW ETHIC

I have now completed the account of the background necessary for assessing the third aspect of "the Frankenstein thing," the moral obligations to the creature itself. I have surveyed the radical changes that are occurring in the social ethic for the treatment of animals and given reasons for their recent rise. I have also spent some time explicating the conceptual content of these changes. With all of this material as a guide, we can critically assess the genetic engineering of animals from the point of view of animal welfare, and we can suggest plausible rules for assuring that such genetic engineering will accord with the ethic described and thus be socially acceptable in this area as well.

Whether one considers the new social ethic for animals a welfare ethic, a welfare ethic with a rights component, or an embryonic rights ethic, one point is clear in practical terms. That ethic is focusing on a far greater range of animal suffering than social morality ever did before. As we saw, this is partly because the twentieth century has witnessed major new sources of suffering, notably accelerated animal research and the use of technological, efficiency-oriented agriculture, both of which need not respect animal natures. The result is that society is concerned with preventing and ameliorating as far as possible the full range of animal suffering created at human hands – not only physical pain, but distress, anxiety, fear, loneliness, boredom, social deprivation, frustration, and so on. Common sense is very comfortable with attributing these modalities to animals, and with identifying them. Part of how they are identified is by comparing the life we allow the animal to live with the sort of life it was evolved (or selected) to lead. When we know that an animal is social in nature and roams over large territories, we consider keeping it alone and in a small cage as inflicting suffering upon it, albeit not necessarily physical pain. On the positive side, common sense sees an animal that is "doing its thing" – fulfilling its nature – as a "happy" animal. As in our ethic for humans, we find it much easier to identify those conditions that lead to suffering and unhappiness than those that definitely create happiness, and thus much of our social concern – indeed, the bulk of our social concern about animals – is with ameliorating suffering.

The fundamental principle that has emerged from our discussions is easy to see: First and foremost, those who are engaged in genetically engineering animals should respect the social demand for controlling pain, suffering, frustration, anxiety, boredom, fear, and other forms of unhappiness or suffering in the animals they manipulate. This is true both for the pilot animals that are studied during basic scientific activity and for the resultant animals that are developed or commercialized on a large scale.

Before developing this point in greater detail, one consideration that is often overlooked, most notably by animal advocates and Rifkin-style opponents of genetic engineering, must be emphasized. There is a tendency on the part of these constituencies to portray genetic engineering of animals as necessarily and universally deleterious to animal well-being. Any such change, it is suggested, will inevitably cause pain and suffering in the animals, and thus they adopt a strong oppositional stance to any genetic manipulation. Conceptually, this position is ill founded, though it does contain, or at least point to, a dimension of concern that is valid, as we shall shortly discuss.

HOW GENETIC ENGINEERING COULD HELP ANIMALS

It is simply false that all genetic engineering must harm animals. Unless one assumes that all species of animals exist currently at their maximal possible state of happiness or well-being or welfare, such a claim is not legitimate. There are numerous examples of cases in which genetic engineering could be used to improve animal welfare. One can genetically engineer disease resistance into animals – this has already been done with chickens and resistance to certain sorts of tumors. One could effect changes in the animal that spare the animal pain and suffering. At the moment, large numbers of cattle are dehorned surgically, without anesthesia, which can be a bloody and painful process. (This is done to keep the animals from harming humans or each other.) There exists, however, a gene for hornlessness, known as the poll gene – there is, for example, a strain of polled (hornless) Herefords. Through genetic engineering, one could introduce the poll gene into all varieties of cattle, thereby obviating the need for dehorning.

Third, one can, in principle, use genetic engineering to deal with genetic diseases in animals. Somewhere around five hundred genetic diseases are found in purebred dogs, for example: blood diseases in Dobermans, bone cancer in Great Danes, eye diseases in collies and shelties; skin diseases in shar-peis. These could be treated either by gene therapy or by actually altering the defective genome. (In gene therapy, or somatic therapy, one in essence delivers the proper genes to the cells of the body so as to provide the gene product whose absence is responsible for the disease. This alleviates the problem in the individual animal or human but does not cure the underlying genetic defect. Alternatively, one can, in principle, at an early embryonic stage of development, replace the defective gene with new genetic material, thereby not only curing that particular human or animal, but also preventing any further transmission of that defective gene to subsequent generations.)

There are various other ways that genetic engineering could be used to augment animal welfare. One could, in principle, engineer food animals to fit environments for which they are not currently suited, thereby allowing them to inhabit areas where they could live in nonconfinement conditions. More radically, one could genetically alter animals to better fit confinement conditions, even to be happy in confinement conditions, if society does not choose to abandon confinement agriculture or radically curtail it, as the Swedish law has done. While it is certainly a poor alternative to alter animals to fit questionable environments, rather than alter the environments to suit the animals, few would deny that an animal that does mesh with a poor environment is better off than one that does not.

There are, in fact, an indefinite number of ways in which genetic engineering could be used in positive ways to benefit animals, both as individuals and as species. As an example of the latter case, one can envision genetically altering members of an endangered species to be more prolific in their reproduction. One could perhaps engineer various sorts of animals to forage in environments that were hitherto inhospitable, thereby promoting extensive, rather than confinement, rearing of food or fiber animals. One could genetically engineer animals to repel the insects that make their lives miserable.

GENETIC ENGINEERING AND ANIMAL TELOS

One important point needs to be stressed here. It will be recalled that, in my discussion of the new ethic for animals, I stressed that it is pivotal to that ethic to protect in animals their most fundamental interests as determined by their natures or telos, even as we do this with humans. This notion has been misunderstood in very serious ways. It has been interpreted by critics of genetic engineering (and sometimes by its supporters) as meaning that my position asserts that an animal's telos cannot be altered without violation, and thus, according to my account, all genetic engineering is wrong. I did not, of course, make that claim. What I did assert was that given an animal's telos, and the interests that are constitutive thereof, one should not violate those interests. I never argued that the telos itself could not be changed. If the animals could be made happier by changing their natures, I see no moral problem in doing so (unless, of course, the changes harm or

171

endanger other animals, humans, or the environment). Telos is not sacred; what is sacred are the interests that follow from it.

Consider a case, which I will also discuss later, where one might indeed be tempted to change the telos of an animal – chickens kept in battery cages for efficient, high-yield egg production. It is now recognized that such a production system frustrates numerous significant aspects of chicken behavior under natural conditions, including nesting behavior, and that frustration of this basic need or drive results in a mode of suffering for the animals.[17] Let us suppose that we have identified the gene or genes that code for the drive to nest. In addition, suppose we can ablate that gene or substitute a gene (probably *per impossibile*) that creates a new kind of chicken, one that achieves satisfaction by laying an egg in a cage. Would that be wrong in terms of the ethic I have described?

If we identify an animal's telos as being genetically based and environmentally expressed, we have now changed the chicken's telos so that the animal that is forced by us to live in a battery cage is satisfying more of its nature than is the animal that still has the gene coding for nesting. Have we done something morally wrong?

I would argue that we have not. Recall that a key feature, perhaps *the* key feature, of the new ethic for animals I have described, is concern for preventing animal suffering and augmenting animal happiness, which I have argued involves satisfaction of telos. I have also argued that the primary, pressing concern is the former, the mitigating of suffering at human hands, given the proliferation of suffering that has occurred in the twentieth century. I have also argued that suffering can be occasioned in many ways, from infliction of physical pain to prevention of satisfying basic drives. So when we engineer the new kind of chicken that prefers laying in a cage and we eliminate the nesting urge, we have removed a source of suffering. Given the animal's changed telos, the new chicken is now suffering less than its predecessor and is thus closer to being happy, that is, satisfying the dictates of its nature.

This account may appear to be open to a possible objection

that is well-known in human ethics. As John Stuart Mill queried in his *Utilitarianism*, "Is it better to be a satisfied pig or a dissatisfied Socrates?" His response, famously inconsistent with his emphasis on pleasure and pain as the only morally relevant dimensions of human life, is that it is better to be a dissatisfied Socrates. In other words, we intuitively consider the solution to human suffering offered, for example, in *Brave New World*, where people do not suffer under bad conditions because they are high on drugs, to be morally reprehensible, even though people feel happy and do not experience suffering. Why then, would we consider genetic manipulation of animals to eliminate the need that is being violated by the conditions under which we keep them to be morally acceptable?

This is an interesting and important objection, amenable to a number of different responses. Let us begin with the *Brave New World* case. Our immediate response to that situation is that the repressive society should be changed to fit humans, rather than our doctoring humans (chemically or genetically) to fit the repressive society. It is, after all, more sensible to alter clothes that do not fit than to perform surgery on the body to make it fit the clothes. And it is certainly possible and plausible to do this. So we blame the *Brave New World* situation for not attacking the problem.

This is similarly the case with the chickens. We know that laying chickens lived happily and produced eggs under conditions where they could nest for millennia. It is our greed that has forced them into an unnatural situation and made them suffer – why should we change them, rather than succumb to greed? This seems to be a simple point of fairness.

A disanalogy between the two cases arises at this point. We do not accept any claim that asserts that human society must be structured so that people are totally miserable unless they are radically altered or their consciousness distorted. Given our historical moral emphasis on reason and autonomy as nonnegotiable ultimate goods for humans, we believe in holding on to them, come what may. Efficiency, productivity, wealth – none of these trump reason and autonomy, and thus the *Brave New World* scenario is deemed unacceptable. On the other hand, were Mill not a product of the same historical values but was rather truly consistent in his concern only for pleasure and pain, the *Brave New World* approach or otherwise changing people to make them feel good would be a perfectly reasonable solution.

In the case of animals, however, there are no ur-values like freedom and reason lurking in the background. We furthermore have a historical tradition as old as domestication for changing (primarily agricultural) animal telos (through artificial selection) to fit animals into human society to serve human needs. We selected for nonaggressive animals, animals that depend on us not only on themselves, animals disinclined or unable to leave our protection, and so on. Our operative concern has always been to fit animals to us with as little friction as possible – as discussed, this assured both success for farmers and good lives for the animals.

If we now consider it essential to raise animals under conditions like battery cages, it is not morally jarring to consider changing their telos to fit those conditions in the way that it jars us to consider changing humans.

Why then does it appear to some people to be prima facie somewhat morally problematic to suggest tampering with the animal's telos to remove suffering? In large part, I believe, because people are not convinced that we can't change the conditions rather than the animal. (Most people are not even aware how far confinement agriculture has moved from traditional agriculture. A large East Coast chicken producer for many years ran television ads showing chickens in a barnyard and alleging that he raised "happy chickens.") If people in general do become aware of how animals are raised, as occurred in Sweden and as animal activists are working to accomplish here, they will doubtless demand, just as the Swedes did, first of all a change in raising conditions, not a change in the animals.

On the other hand, suppose the industry manages to convince the public that we cannot possibly change the conditions under which the animals are raised or that such changes would be outrageously costly to the consumer. And let us further suppose that people still want animal products, rather than choosing a vegetarian lifestyle. There is no reason to believe that people will ignore the suffering of the animals. If changing the animals by genetic engineering is the only way to assure that they do not suffer (the chief concern of the new ethic), people will surely accept that strategy, though doubtless with some reluctance.

From whence would stem such reluctance, and would it be a morally justified reluctance? Some of the reluctance would probably stem from slippery slope concerns – what next? Is the world changing too quickly, slipping out of our grasp? This is a normal human reflexive response to change – people reacted that way to the automobile. The relevant moral dimension is consequentialist; might not such change have results that will cause problems later? Might this not signal other major changes we are not expecting?

Closely related to that is a queasiness that is at root aesthetic. The chicken sitting in a nest is a powerful aesthetic image, analogous to cows grazing in green fields. A chicken without that urge jars us. But when people realize that the choice is between a new variety of chicken, one *without* the urge to nest and denied the opportunity to build a nest by how it is raised, and a traditional chicken *with* the urge to nest that is denied the opportunity to build a nest, and the latter is suffering while the former is not, they will accept the removal of the urge, though they are likelier to be reinforced in their demand for changing the system of rearing and, perhaps, in their willingness to pay for reform of battery cages. This leads directly to my final point.

The most significant justified moral reluctance would probably come from the virtue ethic component of morality discussed earlier in the section on Rifkin reflecting on the unexplained reluctance I have just discussed. Genetically engineering chickens to no longer want to nest could well evoke the following sort of musings: "Is this the sort of solution we are nurturing in society in our emphasis on economic growth, productivity, and efficiency? Are we so unwilling to pay more for things that we do not hesitate to change animals that we have successfully been in a contractual relationship with since the dawn of civilization? Do we really want to encourage a mind-set willing to change venerable and tested aspects of nature at the drop of a hat for the sake of a few pennies? Is tradition of no value?" In the face of this sort of component to moral thought, I suspect that society might well resist the changing of telos. But at the same time, people will be forced to take welfare concerns more seriously and to decide whether they are willing to pay for tradition and amelioration of animal suffering, or whether they will accept the "quick fix" of telos alteration. Again, I suspect that such musings will lead to changes in husbandry, rather than changes in chickens.

THE PRINCIPLE OF CONSERVATION OF WELFARE

The discussion above indicates that genetic engineering of animals is not necessarily deleterious to the well-being of animals. As in the risk areas of genetic engineering of animals discussed earlier, the issues of animal welfare and suffering associated with genetic engineering must be decided on a case-by-case basis. In the welfare area, however, we are not dealing with risks, but inevitabilities – increased or decreased animal suffering. Thus, even if we adjudicated the welfare issues on a case-by-case basis, we still need some overriding ethical principle by which to judge the cases and to provide us with an ideal. And I think this principle is indeed implicit in the previous discussion in this chapter of the emerging social ethic on animal treatment.

Let us recall some salient points about the new social ethic. I have argued that its main source was the dramatic way in which animal use in society changed in the mid–twentieth century. Whereas agriculture was (and still is) overwhelmingly the main user of animals in society, modern technological agriculture as developed in the last fifty years is quite different from traditional agriculture. Whereas traditional agriculture required that animals' biological natures be respected and nurtured, and thus the main source of suffering imposed on animals was a result of cruelty, not of normal use, modern agriculture has "technological fixes" that allow agriculture to override the traditional necessary respect for animals' natures and still be productive and efficient economically. So the contractual, husbandry dimension of agriculture became transformed into something patently exploitative, with significant new animal suffering a by-product of technology. At the same time, other uses of animals developed – notably in research and product testing – that again do not represent a fair contract but are much more exploitative. As society gradually became aware of these new uses, it became increasingly concerned about the proliferation of animal suffering that was not a result of cruelty and was, indeed, associated with activities that benefited society. Nonetheless, society has sought to restrict this suffering and restore at least some of the fairness and respect for animals' natures that characterized our use of animals in traditional animal agriculture. Toward this end, society has begun, in a limited way, to attempt to articulate the rights of animals in law and policy. This is most clearly evidenced by the Swedish law for agricultural animals discussed above, but it can also be seen to some extent in new laws governing animal research, which are aimed at restricting animal pain and suffering.

Thus we may fairly conclude that the new ethic wants animals used by humans to suffer as little as possible. Though it seems to accept some suffering – for example, in the relatively rare cases in medical research in which the suffering cannot possibly be controlled – it demands the control of all pain that can be controlled, regardless of expense or inconvenience. It is also not comfortable with suffering associated with other less exigent uses – cosmetic testing, for example, fur production, or, as European foment has shown, even producing cheap food. The British experience has shown that a significant number of people will pay more for eggs that are produced by free-range, as opposed to caged and confined, chickens. The public no longer tolerates zoos that are perceived as prisons.

It is fair, then, to schematize the primary concern of the new ethic with preventing, mitigating, ameliorating, and redressing suffering and pain that animals used in society experience. So, with genetic engineering, we have the rise of a new technology which, like high-technology confinement agriculture, can make social use of animals more efficient and productive but which in the course of doing so, can also proliferate animal pain and suffering. In a penetrating comment made in open discussion at a 1989 international meeting on farm animal welfare, Professor Stanley Curtis, an animal scientist and farm animal ethologist who often speaks for the swine industry, made the following point: When the agricultural industry developed high technology, high confinement, intensive agriculture – what the public calls factory farming – after World War II, no one was thinking about animals' natures or well-being except as productive units. The only values driving the mechanization of agriculture were concerns about producing cheap and plentiful food for a rapidly growing population. The industry was essentially blindsided by the rise of massive social concern about welfare, he continued, and thus needs time to vector these concerns into its production systems.

Genetic engineering of animals is essentially in the same position vis à vis animals that industrialized agriculture was at its inception. A major difference, however, is that genetic engineering knows of the public concern about animal suffering and, as we shall see later in this chapter, is well aware of both the sort of animal suffering that can be generated by engineering animals without thinking about their welfare and the public moral concern about such suffering. Indeed, critics of genetic engineering such as Jeremy Rifkin and Michael Fox have made this dimension of concern clear and articulate. In fact, many of the genetic engineers with whom I dialogue stress the need for not creating suffering animals through genetic engineering, or else the public is likely to reject the technology. The lesson is manifest that society will not accept new technology that entails new sources of animal suffering when society is in the process of going back to older technologies and rejecting and demanding the control of the suffering those technologies create. Society does want animal use, but it wants animal use to be fair, not one-sidedly exploitative, as a prominent animal scientist candidly remarked in an interview that appeared in an agriculture journal.[18] We have seen that our emerging social ethic in essence demands curtailment of animal suffering even in established areas where humans derive clear benefit – research and factory farming. All animal users are being compelled to step back and clean house, even at the cost of "efficiency." In such a context, in which even time-honored animal suffering is questioned, it is not difficult to see that the clear social maxim for the third aspect of the Frankenstein thing is "restrict new suffering." In other words, genetic engineering should not be used in ways that increase or perpetuate animal suffering. This in turn implies a moral regulatory principle for genetic engineering of animals: Any animals that are genetically engineered for human use or even for environmental benefit should be no worse off, in terms of suffering, after the new traits are introduced into the genome than the parent stock was prior to the insertion of the new genetic material. We may call this the principle of conservation of welfare. In other words, genetic engineering should at least be neutral regarding the well-being of animals. Ideally, it should improve the well-being of the animals engineered over that of the parent stock. Introducing genes for additional disease resistance that do not otherwise harm the organism is a clear example of a genetic modification that meets this test.

Utilizing this principle as a yardstick, how does genetic engineering fare? Here we must hark back to the point I mentioned earlier, that those who say that genetic engineering must harm animals, while conceptually muddled, are raising an important pragmatic point. The point is that the way in which fields like confinement agriculture and animal research have used animals for human benefit and profit has largely ignored considerations of animal well-being, suffering, and happiness. Welfare has been a concern only insofar as it is relevant to the goal of the animal use. Indeed, even when poor welfare has been harmful to the result sought, welfare considerations have not been given top priority by animal users.

Thus, in animal research, it has long been known that "stresses" of various sorts, including pain and fear, have major effects on physiological and metabolic variables in animals relevant to the research uses to which they are put. In other words, if one stresses an animal, changes will occur in the animal that can well color and jeopardize what the research is looking at. D. Gärtner, for example, showed that as putatively trivial a manipulation as moving a cage of rats a few feet can produce dramatic changes in the blood plasma that give a microcirculatory shock profile.[19] Much cancer research has in the past been compromised by keeping animals under conditions so bad that one does not know whether the conditions or the treatment being applied to the animals led to tumor development.[20] Despite the centrality of controlling stress to the research mission, many major research institutions were extremely cavalier about animal husbandry, care, and facilities. This is much improved of late, largely due to the federal legislation discussed earlier.

A similar point can be made about modern animal agriculture. Too often welfare has simply meant enough food and water so the animal can grow or otherwise produce. Even welfare considerations that could increase profit are often ignored. For example, shipping of livestock is clearly unpleasant for the animals – it is hot, crowded, and stressful, and animals excrete on each other. These conditions can result in "shipping fever" and bruising, both of which cost the industry money. Yet little has been done to improve livestock shipping and handling. As a second example, branding of cattle costs the industry almost some $250 million a year in lost hide value, as well as causing much pain – it is, after all, a form of producing a third-degree burn. Yet little is being done to generate an alternative.

These examples indicate that welfare of animals is not a major value for animal-using industries in our era – that is indeed, as we saw, a major reason why the new ethic has emerged in society. Animals are cheap components of large systems. In the high confinement swine industry, for example, each pig generates only a minuscule profit – to survive, the producer must produce high volume. This, in turn, means that little attention can be paid to individual animals. This is equally true in feedlots, broiler facilities, and egg factories. Similarly, in research, for a long time far more concern went into caring for expensive equipment than into animal care – animals were cheap and easily replaced.

Given this manifest tendency, opponents of genetic engineering of animals are right to fear that such engineering will proliferate animal suffering, though they are wrong in thinking that it must do so. In other words, if genetic engineering of animals is allowed to proceed unconstrained, it is likely that concerns for animal welfare, animal suffering, and animal happiness will occupy positions of low priority as compared with considerations of profit and short-run "efficiency." Part of the problem is more global than animal welfare – we, as a society, have become obsessed with efficiency and productivity as the ultimate, sine qua non, values to strive for. We have been seduced by an industrial model that we have indiscriminately applied across the board, even to areas where it is not appropriate. I have already cited the example of universities and the substitution of efficiency for quality.

I also mentioned that similar lack of concern for non–bottom-line values has led to the disappearance of small family farms as "outmoded, inefficient, economic units." Once again, the problem is that other values are eclipsed, such as husbandry, sustainability, preservation of rural communities, and so forth. Only lately have we begun to realize the major environmental and social costs of high-input, large-scale, chemical-dependent, corporate agriculture, which operates with the mind-set I exemplified earlier when it develops herbicide-resistant seeds so we can soak the earth with chemicals despite the existence of a prior problem with groundwater contamination.

In other words, while genetic engineering of animals does not compel the creation of animal suffering, it is likely to lead to it in the current exploitative business context. Thus, for example, if a company discovers a way to genetically engineer chickens so as to get them to produce more eggs, at the cost of additional suffering, it is likely that the company and the industry will perceive this innovation not as a morally questionable step, but as a competitive edge! Such a stance is in perfect harmony with the tendency of industrialized agriculture to push animals as hard as it can to squeeze the most profit and production out of them. If the genetically engineered chicken will economically outperform others, it will be adopted, even if one admitted consequence is increased suffering. A good example of this exploitative approach accomplished by traditional breeding is the turkey, which has been so overbred for a large breast that it cannot mate naturally and must be reproduced by artificial insemination.

Does this, then, mean that the correct response to genetically engineered animals vis à vis animal welfare is blanket rejection? I think not, for by such a tack one loses the good that genetic engineering can do for animals discussed above, as well as the good it can do for humans without harming animals. In any case, as we shall shortly see regarding genetically engineered disease models, it is unlikely that society will reject this technology outright, given certain plausible benefits that it can generate.

The answer to this concern lies not in rejecting genetic engineering of animals totally – that would be like rejecting automobiles because they lead to problems – but in regulating it as regards animal welfare. Enacting in law something like the principle of conservation of welfare enunciated above, mandating that the committees discussed in the last chapter function in accordance with this dictum, and putting the burden of proof for adherence to the principle on those who would genetically engineer a given set of animals would effectively forestall the breakneck development of animals that suffer for profit. At the same time, it would allow for a good deal of development of genetically engineered animals, as a result of which both humans and animals are better off. All such a requirement would do, in essence, is prevent people from taking the easiest path – generating animals by genetic engineering for human benefit regardless of the effect on the animals. It would not make genetic engineering impossible. Rather, it would change the ground rules for genetic engineering. Instead of producing chickens that laid a great many eggs whether or not they suffered additionally, it would require the production of happy chickens that produced a great many eggs.

This is certainly a challenge to genetic engineers, but it fits precisely the ethic that we have seen is emerging anyway. If the biotechnology community argues that my requirement is impossible to meet, it is in essence saying what its opponents argue, namely that genetic engineering does entail creating more animal suffering, and thus it has conceded the issue and irreparably damaged its own case with a society that demands less, not more, animal suffering. Personally, I believe that genetic engineering of animals can proceed as a win-win situation, but only if that requirement is legislated, thereby changing the mind-set that tends to ignore animal welfare. The greatest challenge, as we shall see, lies in the area of biomedicine.

Before looking at welfare in relation to genetically engineered animals in particular areas such as agriculture and biomedicine, one needs to make a fundamental distinction to which scientists are sensitive, but others generally are not. One must distinguish between genetically engineering animals in a research context versus genetically engineering them in large numbers for commercial purposes. Obviously, all genetic engineering will begin with research. Without research, one cannot know whether a particular genetic manipulation will or will not cause animal suffering, simply because, as I pointed out earlier, genetic engineering is far more complex than a one gene, one trait isomorphism. (This is one of the major problems in risk assessment.) Clearly, when one is trying a genetic modification, one cannot know what will happen in advance (as we shall shortly discuss vis à vis the creation of an animal for modeling Lesch-Nyhan's disease). One can sometimes guess, but often one cannot. Thus even a piece of genetic engineering specifically designed to improve animal welfare could have deleterious effects. So the question arises as to what role our principle of conservation of welfare plays in the fundamental research associated with developing the field of genetic engineering itself. (I shall discuss the applicability of the principle to biomedical research shortly.)

The answer is that the principle coincides closely with extant federal research animal legislation. It is designed not to stop research, but to assure that the animals not suffer. Thus, if a local committee were adjudicating a company's proposal to genetically engineer a farm animal to be able to digest forage of a new sort, the committee would ask for detailed descriptions of when the research would be stopped if the animal were not coping as expected, that is, demand criteria for immediate euthanasia or feed supplementation if the animals were not being nourished. It would also demand that the fewest possible number of animals be used; it would require protocols for treating disease; and so forth. This is what current law demands of all animal research. Unfortunately, there are loopholes in the laws that exempt, for example, industrial laboratories doing food and fiber – that is, agricultural research – or research on rodents. If our proposed regulatory mechanism for genetic engineering were to work properly, all animals, when there is reason to believe pain and other sorts of suffering can occur, regardless of species or purpose of the research, would have to be equally considered.

In other words, the basic mechanism that is in place for controlling suffering in experimental animals is applicable to those animals being used to study genetic engineering, or at least it could be rendered applicable with relatively minor modification and expansion. So this aspect of applying our principle of conservation of welfare is not unprecedented. Current law, in fact, even presses (but does not force) researchers to use "lower" animals or nonanimal alternatives wherever possible. Some basic mechanisms of genetic engineering could, for example, be studied in plants and lower invertebrates, entities that are not conscious and cannot suffer.

What is more revolutionary about our proposal of local regulation in accordance with the general principle of conservation of welfare is that it would prohibit, for example, a company from mass-producing and marketing the genetically engineered chicken that produces more eggs but suffers more. This in turn suggests that it would be in the interest of people doing research on genetic engineering of animals, with an eye toward mass production of the resulting animal, both to aim toward producing animals that do not suffer and also to generate data documenting the relationship between the modification in question and the well-being – including pain and suffering – of the animal. In this way research into genetic engineering of animals would also, de facto, be research into animal welfare.

Thus assurance of animal well-being would operate on two levels. The first level would assure that the suffering of animals produced in genetic engineering research, either basic research or commercially oriented research, would be controlled. Federal law already requires this to some extent, but it could be further refined and augmented, as I shall exemplify in my discussion of genetically engineered animals that model human disease.

The second level would essentially assure that no animal whose life violates the principle of conservation of welfare would be produced on a commercial level. This is aimed at prohibiting the production of massive numbers of animals living miserable lives. On the other hand, assurance that the genetically engineered animals were not suffering or were happy would allow production of the animals, provided that the other concerns about risk, discussed earlier, were also met. In this way, concern for the animals is factored into genetic engineering of animals. I can see no other way to deal with this third moral aspect of "the Frankenstein thing."

There is one important philosophical concern about our principle of conservation of welfare that must be addressed. This conceptual problem grows out of a significant philosophical literature concerning our obligations to future generations of humans, but might be applied, mutatis mutandis, to the treatment of animals. The objection is essentially this: Suppose we genetically engineer a new kind of pig, which is defective in some ways, say it has leg and joint problems as a result of being engineered for greater size and more meat, but still lives a tolerable existence, that is, a life of significant but not total misery because of how we engineered it. It could be argued that I have not harmed these new pigs because they are better off than they would have been if I had not genetically interfered with their ancestors, since had I not done so, these pigs would have had no life at all. Thus, according to this argument and contrary to our principle of conservation of welfare, we *are* allowed to create animals that are less happy than the parent stock as long as they are still happy enough to have lives that are worth living.

Though this is a profound and vexatious general point, I believe it can be handled here. In the first place, I am not clear that the evolving social ethic we have discussed is open to that logic. For, by similar reasoning, which has in fact been employed by intensive agriculturalists, animals kept under confinement conditions have, as it were, no complaint, since they owe their lives to these confinement conditions. If society accepted this argument, it would not make the attempt to change confinement agriculture that it is, in fact, making. In terms of the new ethic, an animal having an on balance "tolerable" life is not enough – society wants the animals to have a good life or, at the very least, a better life than life in confinement provides, even if life in confinement is better than no life at all – which, incidentally, many people doubt.

In any case, the logic of the argument is at loggerheads with many other moral concerns that we consider legitimate to address about as yet unborn people and animals. In genetic counseling, for example, we do not worry only about cases where children would be born with intolerable lives; we worry about avoiding suffering that would be a good deal less than intolerable. Similarly, we consider the perpetuation of genetic defects and diseases in purebred dogs for aesthetic reasons embodied in breed standards morally problematic and in need of rectification even though the animals are not totally miserable. (Bulldogs, as I've mentioned, have chronic respiratory problems as a result of our breed standards of dramatically foreshortened physiognomies.[21])

The objection seems to confuse two separate contexts – a context of prospective planning and a context of retrospective blame. Let us recall the remarks by Professor Stanley Curtis that I cited earlier. It may well be that animals raised under confinement since World War II, assuming that their lives are tolerable, have (or their advocates have) no complaint against those who created these confinement systems, since concern about respecting the animals' natures beyond the productive dimensions was not part of the universe of discourse or articulated social values when the systems were created. However, now that these concerns are part of our moral toolbox, we *can* blame those who choose to ignore them and who instead continue to focus only on efficiency and productivity in the design of systems that cause even more suffering. In other words, in the case of genetic engineering of animals, we are now in a position to decide the moral ground rules by which it will be conducted. The social ethic now asks for more than tolerable lives for the animals we use. It asks for happy lives for the animals we raise for food, for example, and the control of pain and suffering, if at all possible, for the animals we use for advancing human health through research, and it is beginning to ask for enriched environments for these animals as well (a step toward happiness beyond the control of pain and suffering.) These are nonnegotiable moral demands that show every indication of increasing, not decreasing, in stringency.

As I have said, society seems to want to continue to use animals for human benefit but wants to do so under the moral constraints just indicated. Therefore, it is reasonable to approach a new technology like genetic engineering in terms of planning for it to accord with these constraints. The principle of conservation of welfare appears to capture in a simple way the constraints that society wishes to impose on both old and new animal use.

GENETIC ENGINEERING AND THE WELFARE OF
AGRICULTURAL ANIMALS

What sorts of things have been done in the area of genetically engineered animals; what sorts of things are reasonable to project; and what are the welfare implications and problems of these activities? Not surprisingly, the areas of greatest activity are animal agriculture and animal research. I shall consider these areas in turn.

The first attempts to employ genetic engineering in an animal agricultural context involved the insertion of the human growth hormone gene into pigs (the so-called Beltsville pigs) and sheep.[22] The desired results were to increase growth rates and weight gain in farm animals, reduce carcass fat, and increase feed efficiency. Although certain of the goals were achieved – in pigs, rate of gain increased by 15 percent, feed efficiency increased by 18 percent, and carcass fat was reduced by 80 percent – unanticipated effects, with significantly negative impact on the animals' well-being, also occurred.[23] Life-shortening pathologic changes including kidney and liver problems were noted in many of the animals. The animals also exhibited a wide variety of diseases and symptoms, including lethargy, lameness, uncoordinated gait, bulging eyes, thickened skin, gastric ulcers, severe synovitis, degenerative joint disease, heart disease of various kinds, nephritis, and pneumonia. Sexual behavior was anomalous – females were anestrous and boars lacked libido. Other problems included tendencies toward diabetes and compromised immune function.[24] The sheep fared better for the first six months but then became unhealthy.[25]

There are certain lessons to be learned from these experiments. In the first place, although similar experiments had been done earlier on mice, mice did not show many of the undesirable side effects. Thus it is difficult to extrapolate in a linear way from species to species when it comes to genetic engineering, even when, on the surface, the same sort of genetic manipulation is being attempted.[26]

Second, as we saw in the chapter on risk, it is impossible to effect simple one-to-one correspondence between gene transfer and the appearance of desired phenotypic traits. Genes may have multiple effects; traits may be under the control of multiple genes. The relevance of this point to welfare is obvious and analogous to a point I made earlier about risk – one should be extremely circumspect in one's engineering until one has a good grasp of the physiological mechanisms affected by a gene or set of genes. One good example of the welfare pitfalls is provided by recent attempts to genetically engineer mice to produce greater amounts of interleukin–4, in order to study certain aspects of the immune system. This, in fact, surprisingly resulted in these animals experiencing osteoporosis, a disease resulting in bone fragility, clearly a welfare problem.[27]

Another example is provided by a recent attempt to produce cattle genetically engineered for double muscling. Though the calf was born showing no apparent problems, within a month it was unable to stand up on its own, for reasons that are not yet clear.[28] To the researchers' credit, the calf was immediately euthanized. Yet another bizarre instance of totally unanticipated welfare problems can be found in the situation where leglessness and craniofacial malformations resulted from the insertion of an apparently totally unrelated gene into mice.[29]

Thus committees should demand that, in pilot research on agricultural animals, a small number of animals be used and that early end points for euthanasia of these animals be established in advance and implemented at the first sign of suffering or problems that lead to suffering, unless such suffering or disease can be medically managed. Such a demand already exists in the laws governing animals used in biomedical research but, as mentioned earlier, these laws do not apply to food and fiber research. Clearly this loophole must be closed before genetic engineering of farm animals, even at an experimental level, is allowed to proceed. Currently, the situation is absurd. One of my colleagues has a colony of research sheep. If a ewe of his gives birth to twin lambs, one lamb may go to biomedical research, the sibling to agricultural research. If both go to experiments requiring castration, the biomedical research lamb must have it done under aseptic conditions, with anesthesia and postoperative analgesia. His unfortunate brother, however, can be treated according to accepted agricultural practice, which could include pocketknife castration or even having the testes bitten off.

The larger problem, however, arises not at the level of research, but at the level of commercial production. Although it is extremely unlikely that either practical constraints or public opinion would allow the Beltsville pig to be commercially viable (recall that it is infertile!), let us imagine that reproduction was not a problem and that the multiple problems experienced by the animal did not impair the increased efficiency and productivity accomplished by gene transfer. In other words, imagine that the pig was economically viable, though suffering in a variety of dimensions. Here we would have a situation analogous to some of the more extreme cases in confinement agriculture, where animals may suffer yet be productive, or where at least the operation as a whole is productive. There is little doubt that these pigs would be commercially produced, given the nature of the industry's bottom line. In confinement agriculture, profits per animal are small; success is possible only through tiny increases in profit magnified over an operation of significant scale. If one could gain a competitive edge through using these animals, they would enter the production arena. For purposes of economic benefit, the suffering and sickness of the animals would not enter the equation!

The true significance of our principle of conservation of welfare thus becomes manifest at the level of commercialization rather than at the level of research. Only by giving the force of federal law to our principle can we prevent defective, suffering, but profitable animals from dominating agriculture. It is not that producers are cruel or insensitive. On the contrary – many producers whom I have dealt with do not like the aspects of intensive agriculture that have forced them to replace animal husbandry with a bottom-line mentality. But it takes only a small number of producers willing to achieve a competitive advantage at the expense of animal welfare to force others to either follow suit or go out of business. As we saw earlier, society is looking for ways to reverse this trend – a good place to start is to prevent the use of genetic engineering to violate animal welfare.

If such a legal constraint were linked to genetic engineering of farm animals, society would have gone a long way toward assuring that genetic engineering of farm animals did not become another force in causing harm to these animals. And all producers would be playing on a level playing field, so the few with no qualms about hurting animals could not gain an edge through their lack of moral concern. (One would also need to prohibit importation of food from countries lacking such laws.) Such a law would not of course stop genetic engineering that benefited both animal and producer, only that which further mined profit at the expense of welfare. In no case can we know in advance, a priori, whether a projected piece of genetic engineering will or will not violate the principle of conservation of welfare. Each must be looked at on a case-by-case basis; hence the regulatory structure I suggested.

We have already encountered some of the plausible strategies for genetically engineering animals that are emerging or are likely to emerge in the near future. Various ways of genetically engineering disease resistance are likely to develop. One way is to introduce genes that code for specific antibodies against pathogens to which the animal in question is not currently resistant. Another strategy is to introduce genes that carry benign portions of viruses that will occupy receptors within the animal so that the pathogenic virus will have, as it were, no place to land.[30] Still another transgenic strategy will doubtless focus on refining growth hormone gene transfer to eliminate the problems I have mentioned while preserving the benefits. If these problems were eliminated, it might be possible to preserve welfare while augmenting productivity.[31]

Experiments are currently under way to increase wool production in sheep.[32] Low availability of the amino acid cysteine can slow wool growth. It is theoretically possible to transfer genes to sheep that would increase synthesis of cysteine and thereby increase the rate of wool growth. Genetic engineering is also likely to be used, as it has already been used in fish, to increase animals' tolerance of cold. Similar attempts will probably be made with heat tolerance as well. It is too early to tell what effects such changes will have on welfare. Monitoring of such work by the mechanism suggested, however, would assure that welfare considerations remained in the hopper.

It is also not inconceivable that as agriculture becomes more responsive to social pressure regarding confinement, it will seize upon genetic engineering as a strategy for better fitting animals to their environments in order to reduce suffering. I do not, however, consider this likely in the foreseeable future. Given the embryonic state of knowledge of the connections between genes and behavior; indeed, given the similarly limited knowledge of behavior, pain, suffering, and so on in animals; and given the extreme likelihood that psychological dimensions relevant to welfare, happiness, and unhappiness are related to a multiplicity of genes, I do not think we can look forward to the day when we will produce chickens that are ecstatic in battery cages.

Suppose, however, this were possible – would it be morally wrong to do so? Obviously, it would be much more simple and reasonable to change the husbandry systems to fit the animals than to change the animals to fit the systems. If, however, ex hypothesi there were only two choices – either leave the animals as they are now, to live under conditions that do not meet their needs, or change their needs so they no longer suffer from the frustration of their fundamental urges – it seems clear that changing the animals is the lesser of the two evils. This, of course, assumes that we would be wise enough to change the animals in such a way as to indeed accommodate their interests without creating new interests that were thwarted or otherwise generating new suffering. But if we could do this, why would it be wrong? At least the animals would be happy, or closer to happy, or at worst not suffering. Admittedly it would be repulsive to do so, but I believe that is an aesthetic revulsion, not necessarily a moral one. A moral component would enter into the discussion only if there were hidden costs to the animals, or if such modifications made us more prone to treat animals merely as tools for human use rather than as "ends in themselves," whose fates matter to them as well as to us.

In my view, caring enough about the animals to try to engineer them so as to be happy, even if ultimately an exploitative stance, is still an improvement over our current mind-set, which develops animals to suit our desires, and, in the case of pets, our whims, with total disregard for the effect on the animals – witness our genetically diseased companion animals, or our surgical mutilations of these beings: ear cropping, tail docking, and so on. In the same vein, if we could genetically engineer essentially decerebrate food animals, animals that have merely a vegetative life but no experiences, I believe it would be better to do this than to put conscious beings into environments in which they are miserable, though again this seems aesthetically abhorrent to us. Perhaps our distaste grows out of the fact that such suggestions force us to recollect, in Plato's sense, how far we have moved from the traditional husbandry contract with animals upon which so much of human history was based. Such activities would show us up as mean-spirited, self-serving, exploitative beings, in the same way that producing happy, idiotic, Stepford-wife robots would. Perhaps the robots could be built to be happy; we still do not like the sort of person who would prefer a happy idiot to another human with a mind and personality of its own with whom we relate in a mode other than an exploitative one.

One genetic engineer has argued that, in the long run, biotechnology will make the whole debate about agricultural animal welfare moot. Eventually, he told me, we will be able to create the relevant animal proteins in fermentation vats produced by bacteria genetically coded to do so. Thus we will be able to have animal products without animals. If this is the case, the welfare issues will indeed disappear.

The one thing we emphatically do not wish to have occur by the genetic engineering of farm animals is the creation of traits that make animals more dependent on confinement while making them no happier in confinement. Fox has pointed out that genetically engineering pigs for little or no body fat would in essence make these animals dependent on being raised in confinement, as they would be unable to cope with cold.[33] Such a genetic alteration, however, would be excluded by our principle of conservation of welfare, for a pig that cannot adjust to temperature change is surely worse off than one that can.

ALTERING ANIMAL PRODUCTS BY GENETIC ENGINEERING

In any event, let us return to less speculative aspects of genetic engineering of animals and continue to look at modifications of animals likely to occur in the near future and the relationship of those modifications to animal welfare. One use that spans both agriculture and biomedicine is the genetic engineering of animals in order to alter their products.

In an agricultural context, one of the most plausible uses of genetic engineering would be to change the composition of milk by genetic engineering. Seidel points out that by genetically engineering changes in casein content of milk, one could increase cheese yield significantly enough to save the dairy industry $190 million per year. One could also modify casein genetically to cut down the time required for cheese ripening or to increase the thermal stability of milk. One could cut fat content of milk genetically or remove lactose genetically, thereby making milk available to large numbers of lactose-intolerant people.[34]

Seidel points out that experiments are under way not only to create true transgenic animals to express and pass on these traits to their offspring, but also to alter animals by somatic therapy; or gene insertion into tissue but not germ cells, so that animals can produce the desired product in the relevant tissues, for example, the mammary glands, for a period of time.[35] This can be accomplished much as is done in human gene therapy – by injection or other physical delivery of the desired DNA into the relevant cells.

Genetic engineering for new products has already occurred

in the biomedical arena. Here work has been done to engineer animals to secrete valuable, biologically active compounds for pharmaceuticals in milk, blood, or any other tissue, for example, egg white. Investigators have succeeded in utilizing this approach to produce TPA (tissue plasminogen activator), a drug used to dissolve blood clots. Other blood-borne coagulation and anticoagulation factors have been successfully produced in the mammary glands of goats, pigs, and sheep. Human hemoglobin has been produced in the blood of pigs. All of this is increasingly important as availability of human blood decreases. Such genetic engineering allows for a far less costly method of extracting these products than those currently employed.[36] Seidel has pointed out that one transgenic goat could supply all the FSH (follicle-stimulating hormone) needed in the world in its milk; FSH is currently extracted from slaughterhouse pituitary glands.[37] One could also genetically alter animals' milk to improve the health of both humans and animals who consume the milk.

In any event, such "molecular pharming" of animals, as it has excusably been called, does not seem to raise dramatic issues for animal welfare, assuming that there are no untoward side effects growing out of the transferred genes. (What a genetic engineer considers a "side effect" can, of course, be a major concern to the animal.) Such animals, being extremely valuable, will presumably be well cared for. My only concern is that they not be kept in isolated, sterile, sanitized environments with no opportunity to fulfill their behavioral and psychological needs. But this is a problem common to most laboratory animals and most farm animals; it should be and has begun to be of concern to society in general. But it is not a special concern vis à vis these genetically engineered animals, growing out of the transgenic technology.

GENETIC ENGINEERING AND THE WELFARE OF RESEARCH ANIMALS

We turn now to the biomedical-scientific uses of transgenic animals. Here, as we shall see, we will encounter some major challenges to animal well-being that go beyond those found in agriculture.

Thus far, the major uses of transgenic animals have been in the area of biomedical research. During the early 1980s, only a few dozen papers were published dealing with research on transgenic animals; as of 1992, the number of such publications exceeded twenty-five hundred.[38] (As this book goes to press, in late 1994, my colleague George Seidel estimates that the number exceeds five thousand.) Thus far, for reasons of cost and ease of care, most transgenic research animals have been transgenic mice, of which more than a thousand strains have been produced. The NIH has created a transgenic mouse facility, the National Transgenic Development Facility, in Princeton, New Jersey, to help make transgenic mice more easily affordable.[39] The program is funded to subsidize annually approximately 100 different efforts at producing transgenic mice. Other transgenic animals that have been developed include amphibians, rats, rabbits, sheep, goats, cattle, swine, poultry, and fish, many of which we have encountered in our discussions. Clearly, given the power of the technology for research, such research will grow exponentially, and it is far better to have the welfare issues clearly articulated in advance – and strategies developed to deal with them – than to have them surface in a sensationalistic context.

Transgenic animals have been used in both pure and applied research. One fundamental area studied through transgenic animal use is developmental biology.[40] Such study is usually accomplished by one of two different strategies – either by inserting new genetic material into the animal, or by ablating endogenous genetic material. By using the first strategy, inserting new genetic material, one can study how different cell lines develop in embryos, which do not ordinarily lend themselves readily to experimental manipulation. By inserting foreign DNA into the embryo, one can observe the fate of that DNA during ensuing development. Foreign genes can also serve as markers for chromosomal regions. One can also study the effects of various genetic contexts in development, for example, development in males versus females.[41] Furthermore, depending on where the foreign gene incorporates into the embryo, it will show different developmental effects.

The second strategy for studying development is that of ablation (removal) of genetic material. This is accomplished by either (1) insertion of toxic genes that are tissue specific, that is, have an effect by killing certain cell lines,[42] or, more commonly, 2) making normal genes inactive so they do not function at all. Many of these experiments are performed on the developing embryo, and they do not raise issues of animal pain, suffering, and well-being because they are terminated before the animal reaches a sufficient level of development to be conscious. Further, given the current state of knowledge, radical alterations induced in the developing embryo would virtually assure that the animal would not survive birth. Exceptions to this rule, or advance of the science and technology to the point where we could produce major monsters that suffer, are covered by our principle of conservation of welfare and would thus be disallowed.

This case shows, by the way, that our criterion is prima facie more strict than current regulations, which allow for the production of such animals provided everything possible is done to control pain and suffering. Requiring termination of such experiments prior to their causing pain and suffering would, I believe, both accord with the new ethic and still allow for much significant study of development. Even if our criterion were not legally mandated, it would behoove scientists to voluntarily design their experiments to eliminate pain and suffering so as not to evoke a massive public rejection of transgenic animals growing out of this third aspect of "the Frankenstein thing."

A similar point can be made regarding a second major area of research that uses transgenic animals, namely, immunological research. Transgenic animals have especially been utilized to study how the immune system recognizes "self" (cells of the organism belonging to the organism) and distinguishes it from nonself (foreign pathogens, transplants, and so forth). Once again, the points I made above apply. Such research should be conducted so as to eliminate any pain and suffering by use of specific end points for euthanizing the animal prior to the advent of suffering; anesthetics, and so on. Again, scientists will doubtless object to a criterion of welfare more stringent than the ones currently in use. If one can currently legally study phenomenon X in animals without the use of genetically engineered animals and not be forced to control all pain and suffering in those animals, why should the use of genetically engineered animals for the same purpose call forth stricter demands? This is a legitimate and vexatious point that must be answered.

My response would be twofold: First of all, I think that the criterion of eliminating all pain and suffering (except for momentary pain of injection, blood drawing, and so on) in animal research is ultimately where the social ethic is going. Many scientists recognize this and are, in fact, turning many of their projects into terminal experiments, which involve no suffering. Second, as has been discussed, social acceptance of genetic engineering is highly tenuous and tentative. A failure to assure society that genetic engineering of animals will not proliferate animal suffering could lead to society's rejecting or significantly impeding all of genetic engineering of animals. An unambiguous rejection of animal suffering regarding this technology, or at least an assurance that suffering *will* be controlled – not just "if possible" – will remove a major obstacle to social trust in genetic engineering of animals and will also assure the public that the scientific community is sensitive to concerns about animals. If the research community does accept such a constraint, it would clearly reflect back onto other areas of animal research and ideally accelerate the rejection of any suffering there as well.

There are other reasons why adopting a stricter criterion for genetically engineered animals is not irrational. As we shall see, failure to adopt such a criterion can lead to a proliferation of animal suffering many orders of magnitude greater than what we have seen before, because the technology, in principle, allows us to create animals that "model" some of the most horrible diseases that affect humans, for which no animal analogues have existed. Such a potential for increasing suffering demands stronger measures. Second, this technology is emerging at exactly the time in history when society is most sensitive to animal suffering and demands its control, even when society benefits from that suffering.

The same sort of comments I made about other areas of research utilizing transgenic animals also apply to a third major area in which they have hitherto been employed – the study of oncogenesis (cancer development). Indeed, the very first animal ever patented in the United States was the Harvard-Dupont mouse disposed to the development of tumors. In the words of the patent, this is "an animal whose germ cells and somatic cells contain an activated oncogene sequence introduced into the animal . . . which increases the probability of the development of neoplasms (particularly malignant tumors) in the animal."[43] In the face of the new laws, the research community has made major advances in minimizing the pain and suffering of animals used in cancer research, advances that apply to such transgenic animals as the Harvard-Dupont mouse. One establishes, for example, end points for animal euthanasia in terms of tumor size, so that the animal is killed painlessly before the tumor can have symptomatic effects. One utilizes anesthetics, analgesics, and tranquilizers in the course of and after the operative procedure. In any case, it is now known that both animals and humans heal and recover better if they do not experience pain and suffering, contrary to both scientific ideology, which treated animal pain as unreal, and certain aspects of medical ideology, which suggested that pain "builds character."[44]

A fourth research use for transgenic animals is the testing and study of approaches to somatic gene therapy, that is, the correction of genetic defects by applying gene transfer technology to somatic (body) cells of the genetically deficient organism. (Strictly speaking, such animals are not true transgenics, as the transferred genes are not incorporated in the recipient animals' germ cells, but only into somatic cells. I include this category because previous discussions of transgenic animals in research have done so.[45]) Since there are over three thousand known genetic diseases, whose effects are often devastating and for which current therapies are often inadequate, a great deal of interest exists in exploring gene therapy. Inbred mice having an abnormal β-globin gene resulting in β-thalassemia, a hemoglobin disease, have been used to test the efficacy of transferring the human β-globin gene as a therapeutic modality.[46] The gene transfer was successful, and the animals were able to produce functional hemoglobin; indeed, they were able to do so more successfully than the normal mouse. Recently, a similar protocol has been demonstrated for humans. Gene therapy in mice has been employed to correct dwarfism, by somatic introduction of growth hormone genes, and to cure "shiverer mice" who suffer from a lack of myelin basic protein.[47] Such uses can again minimize pain and suffering by proper end points and proper pain control protocols. In successful cases, of course, the animals will be better off as a result of the experimental therapy, since the point of such experiments is to alleviate existent symptoms.

CREATING ANIMAL MODELS OF HUMAN GENETIC DISEASE: A DILEMMA

All of the above uses of transgenic animals in research raise welfare problems that are essentially similar to problems that already exist vis à vis nontransgenic laboratory animals, and our responses to those problems have been similar to the responses federal law demands of researchers, except that we have used a stricter criterion. The full value of the stricter criterion can only be understood when we consider a final category of research use of transgenic animals. This is the area of creation and maintenance of seriously defective animals that are developed and propagated to model some human disease. This was traditionally accomplished through identification of adventitious mutations and selective breeding. Transgenic technology allows for accomplishing the same goal far more quickly and in a far broader range of areas. One can, in principle, essentially replicate any human genetic disease in animals, including those that did not exist in animals in the past.

Indeed, a number of new animals genetically engineered to

model human diseases for which there had been no previous animal "models" have already been created. One, the Lesch-Nyhan's mouse, will be discussed in detail shortly. Another is a mouse created to replicate Gaucher's disease, a lysosomal storage disease affecting chiefly Eastern European Jews.[48] Still another group of researchers created rat models for a range of genetic diseases known as spondyloarthropathies, which include ankylosing spondilitis, various forms of arthritis, and inflammatory diseases of the skin, heart, eye, gastrointestinal tract, and genitourinary tract.[49]

Additionally, as we saw in discussing the AIDS mouse earlier, transgenic technology allows us to create animals that can be affected by other human diseases they hitherto could not contract. These new capabilities afforded by genetic engineering have major implications for the third aspect of "the Frankenstein thing," the well-being and suffering of animals, for they issue in the possibility for massive amounts of suffering in the animals created as a vehicle for studying these diseases. A recent chapter in a book devoted to transgenic animals helps to focus the concern:

> There are over 3,000 known genetic diseases. The medical costs as well as the social and emotional costs of genetic disease are enormous. Monogenic disease [i.e., disease controlled by a single gene] accounts for 10% of all admissions to pediatric hospitals in North America . . . and 8.5% of all pediatric deaths. . . . They affect 1% of all liveborn infants . . . and they cause 7% of stillbirths and neonatal deaths. . . . Those survivors with genetic diseases frequently have significant physical, developmental, or social impairment. . . . At present, medical intervention provides complete relief in only about 12% of Mendelian single-gene diseases; in nearly half of all cases, attempts at therapy provide no help at all.[50]

This is the context in which one needs to think about the animal welfare issues growing out of a dilemma associated with transgenic animals in biomedical research. On the one hand, it is clear that researchers will embrace the creation of animal models of human genetic disease as soon as it is technically feasible to do so. Such models, which presumably introduce the defective human genetic machinery into the animal genome, appear to researchers to provide convenient, inexpensive, and – most important – high-fidelity models for the study of the gruesome panoply of human genetic disease outlined in the over three thousand pages of text that constitute the sixth edition of the standard work on genetic disease, *The Metabolic Basis of Inherited Disease*.[51] Such "high-fidelity models" may well reduce the numbers of animals used imprecisely in such research, a major consideration for animal welfare. On the other hand, the creation of such animals will surely generate inestimable amounts of pain and suffering for these animals, since genetic diseases, as mentioned above, often involve symptoms of great severity. Furthermore, the total number of animals used will probably increase, for all human genetic diseases and defects could be replicated in animals. The obvious question then becomes the following: Given that such animals will surely be developed wherever possible for the full range of human genetic disease, how can one assure that vast numbers of these animals do not live lives of constant pain and distress? Such a concern is directly in keeping with the emerging social ethic for the treatment of animals; as I have said, one can plausibly argue that eliminating animal pain and distress is the core of new social concern about animal use.

The first attempt to produce an animal "model" for human genetic disease by transgenic means was the development of a mouse that was designed to replicate Lesch-Nyhan's disease, or hypoxanthine-guanine phosphororibosyltransferase (HPRT) deficiency.[52] Lesch-Nyhan's disease is a particularly horrible genetic disease, leading to a "devastating and untreatable neurologic and behavioral disorder."[53] Patients rarely live beyond their third decade and suffer from spasticity, mental retardation, and choreoathetosis. The most unforgettable and striking aspect of the disease, however, is an irresistible compulsion to self-mutilate, usually manifesting itself as biting fingers and lips. The following clinical description conveys the terrible nature of the disease:

The most striking neurologic feature of the Lesch-Nyhan syndrome is compulsive self-destructive behavior. Between 2 and 16 years of age, affected children begin to bite their fingers, lips and buccal mucosa. This compulsion for self-mutilation becomes so extreme that it may be necessary to keep the elbows in extension with splints, or to wrap the hands with gauze or restrain them in some other manner. In several patients mutilation of lips could only be controlled by extraction of teeth.

The compulsive urge to inflict painful wounds appears to grip the patient irresistibly. Often he will be content until one begins to remove an arm splint. At this point a communicative patient will plead that the restraints be left alone. If one continues in freeing the arm, the patient will become extremely agitated and upset. Finally, when completely unrestrained, he will begin to put the fingers into his mouth. An older patient will plead for help, and if one then takes hold of the arm that has previously been freed, the patient will show obvious relief. If help is not forthcoming, a painful and often severe injury may by inflicted. The apparent urge to bite fingers is often not symmetrical. In many patients it is possible to leave one arm unrestrained without concern, even though freeing the other would result in an immediate attempt at self-mutilation.

These patients also attempt to injure themselves in other ways, by hitting their heads against inanimate objects or by placing their extremities in dangerous places, such as in between the spokes of a wheelchair. If the hands are unrestrained, their mutilation becomes the patient's main concern, and effort to inflict injury in some other manner seems to be sublimated.[54]

As of the present, "there is no effective therapy for the neurologic complications of the Lesch-Nyhan's syndrome."[55] Thus Kelley and Wyngaarden, in their chapter on HPRT deficiency diseases, boldly suggest that "the preferred form of therapy for complete HPRT deficiency (Lesch-Nyhan's syndrome) at the present time is prevention," that is, "therapeutic abortion."[56]

Researchers have sought animal "models" for this syndrome for decades and have, in fact, created rats and monkeys that will self-mutilate by administration of caffeine and other drugs.[57] It is thus not surprising that, as mentioned above, the first disease genetically engineered by embryonic stem cell technology was Lesch-Nyhan's disease.[58] But although the gene was successfully inserted, these animals were phenotypically normal and displayed none of the metabolic or neurologic symptoms characteristic of the disease in humans. The reasons for this are unknown.[59]

This case provides us with an interesting context for our animal welfare discussion. Although the animals were in fact asymptomatic, presumably at some point in the future researchers will almost certainly be able to generate a symptomatic model transgenically. Let us at least assume that this can occur – if it cannot, there is no animal welfare issue to concern us! Given that researchers will certainly generate such animals as quickly as they are able to do so, how can one assure that the animals live lives that are not characterized by the same pain and distress they are created to model?

Once again, this question does not differ in kind from the moral questions associated with developing traditional chronic animal models of human disease, be it by breeding, pharmacological manipulation, or tissue destruction. The difference is in degree; transgenics provides the potential for generating vast numbers of animals modeling genetic disease and other diseases with devastating symptoms.

Here we are faced with a true dilemma. On the one hand, I do not believe that the social ethic will forbid the creation of such animals, since the potential benefit to humans growing out of such animals appears direct and significant. On the other hand, there is no way to study these diseases in acute or terminal or short-term experiments. Lesch-Nyhan's patients, for example, do not show symptoms from birth, but they do exhibit them chronically after their later onset. Furthermore, researchers will certainly wish to observe the whole course of the disease. In other words, they will want to keep the animals alive for long periods of time. And anesthesia cannot be maintained in an animal for periods of months or even weeks. Although one can use analgesics or sedatives chronically, these do not control the compulsion to self-mutilate in humans and thus are unlikely to do so in animals.

What is therefore needed, to *assure* that the animals are not suffering, is not something that changes subjective experience, but something that eliminates it; something akin to general anesthesia that, unlike general anesthesia, can be utilized for the life of the animal. Only in this way could we meet our principle of conservation of welfare, for the animal would be insensitive to what happens to it.

I would therefore argue that the only way to pass between the horns of the dilemma regarding chronically defective, suffering, genetically engineered animals – to stay in harmony with the principle of conservation of welfare while creating such animals – would be to obliterate all subjective experience, to totally eliminate consciousness. One such approach could involve surgically rendering an animal decerebrate, so that, while vegetative functions are extant, the animal's thought and feeling are shut down. Since genetic diseases are metabolic diseases and are not under the control of the cerebral cortex, they could still be studied. The animal would be mentally dead but physically alive. Alternatively, one could perhaps induce coma in the animals, for example, by anesthetic overdose. As a third possibility, one could perhaps genetically engineer these animals both to be Lesch-Nyhan's models and to be born decerebrate – decerebrate progeny are regularly born in many species, including humans.

Radical though these approaches may be, they do not violate the principle of conservation of welfare, since an animal with no mentation or feeling has no welfare, or, if it does, has welfare only in the trivial sense that a plant does. Only in this way can we prevent the creation of creatures whose entire lives are suffering – we would essentially be euthanizing them but keeping them vegetatively alive. If we are correct in our assumption that the capacity for genetically engineering models for all manner of genetic and other diseases is imminent, and that the research community will forge ahead in creating such models, the sort of approach I have outlined seems to be the only viable way to control suffering. Even if my conceptual point is accepted, it is by no means clear that these measures are practicable. Thus inquiry into controlling long-term suffering should be an immediate research priority for the biomedical research community. Those agencies prepared to fund research into creating such animals should also be funding research into controlling their suffering. Unfortunately, given our track record, it is likely that such animals will be created and, only then, will the issue of the unacceptability of their suffering gradually surface, after countless animals have lived miserable lives.

In sum, as regards the third aspect of "the Frankenstein thing," genetic engineering of animals is not necessarily connected with animal suffering. On the other hand, given the fact that both the agricultural community and the research community have, during the past fifty years, essentially ignored issues of animal ethics and welfare, it is doubtful that they will suddenly be able to vector this consideration into their activities vis à vis genetically engineered animals without direction from society. Society as a whole has indeed begun to develop that direction, in the form of what we have called the "new ethic for animals." Since there is no reason to believe that that ethic will diminish and, as we saw, every reason to believe that it will continue to grow, genetic engineering of animals should be regulated in accordance with that ethic. As we saw also, this new ethic does not seek to abandon animal use, but demands assurance that animals used for human benefit do not suffer and are happy. (New laws in the United States and elsewhere attest to the power of this ethic.) As a principle for regulating genetic engineering, I have proposed the principle of conservation of welfare, whereby animals should be no worse off if genetically engineered than they would be if they were not so engineered. Such a rule, encoded in law, will assure not only the well-being of genetically engineered animals but also that genetic engineering will not be at loggerheads with social thought. We have also seen that the greatest challenge lies in the area of using genetic engineering to create models of genetic and other disease.

ANIMAL PATENTING

Much of the controversy surrounding the genetic engineering of animals has been focused on the issue of animal patenting – should those who genetically engineer new animals, such as the Harvard mouse, be allowed to have exclusive ownership of that form of life, exactly as one owns rights to a mechanical or electrical invention? As I mentioned, the U.S. Patent Office essentially made the decision in 1987, and thus far few people consider the issue resolved. Religious groups argued that such patenting violates God's law; environmentalists argued that it would lead to environmental despoliation; farmers feared that it would put small family farms at a disadvantage; animal welfare advocates said it would augment animal suffering. Advocates argued that it would enable the United States to compete better and that patenting animals was innocuous.

In my view, the Patent Office rushed in where angels fear to tread. The issue of patenting is logically dependent on social resolution of the multifarious ethical and prudential issues related to genetic engineering that I have considered in this book. The issuing of patents begs these questions or ignores them. It was a bureaucratic decision made in a value-free context (or value-ignoring context) by an agency that has notoriously avoided engaging the ethical and social issues raised by inventions like switchblades, assault rifles, shock-collars, and devices for sadomasochists, an agency that judges applications only by the formal criteria of novelty, usefulness, and nonobviousness. It disavows concern with issues of safety; danger to humans, animals, or environment; or welfare of animals. The decision is, as it were, a punchline without a joke, an ending without a story. The decision to patent or not to patent should follow in the wake of democratic social examination of the concerns discussed here, and in the wake of establishing a democratic regulatory mechanism for all aspects of genetic engineering of animals. In the case of a technology with far-reaching consequences, we should decide how we as a society wish to deal with it before wholesale commercial exploitation proceeds, for our sakes, the animals' sakes, and the investors' sakes.

As I said at the beginning of this book, bad ethical issues drive good ethical issues out of circulation, and the debate over the consummatory act of patenting provides a good example of that principle. We are arguing about the icing before we have baked the cake.

In my analysis, I have provided discussions and responses to many of the issues upon which the patenting decision should rest. It is simply incoherent to patent organisms without society indicating what risks it will accept regarding such organisms, what benefits it expects, and, most importantly, what mechanism of oversight and regulation it wishes to see in place. Equally, in a society where moral concern for animal treatment is a rapidly growing thrust, it is not sensible to license ownership of animal kinds without a mechanism that lays out the rules for what forms those animals should be allowed to take and how they will be treated. Will society accept a proliferation of defective disease-modeling animals who live lives of constant suffering for human benefit? Will it accept, to take an example actually suggested by a genetic engineer, a ten-thousand-pound cow? [60] Will it accept animals that are nakedly designed to be nothing but factories at a time when it is rejecting, or at least questioning, confinement agriculture?

In other words, the real issue is what rules of welfare will constrain and determine the sorts of animals produced by genetic engineering. At one extreme, society is likely to accept goats that are engineered to produce valuable drugs in their milk, are well cared for, and suffer no adverse effects. At the other, I strongly doubt society will accept Beltsville pigs that suffer in a variety of ways, even if it means cheaper pork, or legless, wingless, featherless chickens who efficiently convert feed into meat. As I have argued, society will probably accept animal models for genetic disease provided a mechanism is in place to assure that they do not suffer. Perhaps my intuitions are wrong; but in any case, the general requirements for the welfare and well-being of genetically engineered animals –

something like my principle of conservation of welfare – should be articulated in advance, so those who genetically engineer animals know the rules of the game, not after the fact.

In my view, the issue of whether humans should own kinds of animals, and whether such ownership violates divine will, is a red herring and an example of my moral Gresham's law. The fact is, in our current legal system, animals are property – domestic animals are the property of individuals, "wild" animals are community property. If all animals are owned, it is hard to see why kinds of animals should not logically be owned, since, as we saw, kinds are groups of individuals. The real issue is not the technical notion of ownership, but rather what constraints are imposed by society on those who own animals. (In Europe, where patenting has been allowed, certain patent applications have been rejected on welfare grounds.) As I once argued in a brief to the Canadian Law Reform Commission, there is legal precedent for owned entities having significant rights: Slaves were property, but even in Roman law one could not treat slaves in any way one wished, and slave murder could in legal theory be severely punished.

Whether or not animals remain property in the eyes of the law, society, in virtue of its new ethic for animals, will demand ever-increasing constraints on how they are treated. In my own view, animals should not legally be property, and the arguments that applied to humans not being owned apply, mutatis mutandis, to animals. Animals should perhaps enjoy a legal status more like that of children, although, unfortunately, whatever their de jure status, children are de facto a lot like animals and are more often than not treated as parental property (though this, too, is slowly changing). The mere fact of ownership, or lack thereof, does not respectively assure bad or good treatment. If strict welfare rules or laws are promulgated and enforced, patented animals could be happy animals. It they are not, then even legal liberation of animals from being property will have no actual meaning as regards their treatment.

Who will be seeking to patent animals? No doubt it will be those large corporate entities that have sufficient capital to invest in such a long and risky process – probably pharmaceutical companies, chemical companies, and the other interests that have industrialized agriculture or have been involved in animal research in a significant way, and have profited significantly from both activities by emphasizing efficiency and productivity.

I want to stress that I am not opposed to profits or corporations – both are integral to our society. In the area of animal welfare, however, corporations have not displayed a leadership role. Their emphasis on squeezing the most from animal agriculture has led to a confinement agriculture based on drugs and chemicals. This industrialized agriculture in turn gives rise to animal welfare problems, environmental problems (e.g., management of waste, groundwater contamination), and concerns about the effects of chemicals on the safety of food (e.g., the concern that widespread use of antibiotics in animal feeds led to selection for antibiotic-resistant pathogens). Furthermore, the industrialization of agriculture has led inexorably to the demise of small farmers, since they lack the capital to compete "efficiently." Most chicken, for example, is produced by a few big companies – the entire industry is vertically integrated (owned top to bottom by the same company), and small producers are essentially serfs to the company. Experts affirm that the pork industry is rapidly moving in the same direction. In addition, highly industrialized agriculture requires vast inputs of energy, fuel, chemicals, and so forth and does not weigh sustainability heavily as a value.

In short, these large corporate entities do not deeply reflect the values of society as a whole – they have done well because they have produced cheap food, with the hidden costs of this food only now becoming apparent to society in general, which seems to have growing reservations about these costs. For example, society wants better animal welfare, more sustainable agriculture, less reliance on drugs and chemicals.

There is, as yet, no evidence that corporate agriculture has assimilated these concerns.

So there is every reason to believe that the animals that such entities develop for patenting are based on the current model for corporate agriculture – high production favoring ever larger and more efficient operations. BST (bovine somatotropin) emerged from this mind-set – indeed, BST is a good case illustrating corporate blindness to social values. As I have already detailed, the public did not welcome BST with open arms. The animals produced through this mind-set are likely to replicate and amplify the problems I have identified. A good example is given by Michael Fox, when he cites a prediction by a leading researcher that, within the next ten to twenty years, industry could produce ten-thousand-pound cattle and pigs twelve feet long and five feet high.[61] To his credit, the researcher points out that this possibility raises a host of problems, including economic, environmental, and ethical ones.

It is not difficult to see how this could be the case. We all know that there are limits on the size of land animals, because structural support does not increase proportionally to size. Humans who are giants suffer limb and joint problems, as did the Beltsville pigs and super-sized chickens developed by a pharmaceutical company. The research animals discussed would, a fortiori, have similar problems, leading to primary suffering in limbs and joints and probably secondary suffering as well, growing out of attempts to compensate for the limb problems. They might well also have problems lying down and getting up. Their reluctance to exercise might also lead to digestive and respiratory problems, and so on. Slaughter of such animals might also present welfare problems, since methods of stunning have been devised for much smaller animals. Transport and handling of such animals would be difficult and perhaps dangerous to personnel as well as to animals. Reproduction by natural means might be difficult.

Environmental problems are also plausible to envision. Assuming that the cattle would be range fed for part of their lives, their vast weight meeting the ground through relatively small hooves would contribute to soil compaction and range degradation. Additionally, if oversized animals were indeed unhealthy and in constant pain, they would be more prone to infection and various other stress-related diseases. This could in turn require more drugs for their management, which could lead to problems of residues, food safety, and so on.

If patented, it is unlikely that small operators could afford such animals, and these farmers, already hanging on by a thread, would be forced out by their inability to compete. (A similar concern exists among researchers regarding patented research animals – only the wealthiest researchers could afford them.) Although some agricultural economists have argued that this only proves that small operations are dinosaurs – "outmoded economic units," as one such economist put it to me – this is a self-serving characterization. The fact that a giant multi-million-dollar diversified corporation – an AT&T or an IBM – can afford to invest huge amounts in new technology that will take a long time to return a profit, since they are making large amounts of money in other areas, while a fourth-generation family rancher working three jobs to keep the ranch going and pay the taxes cannot afford such an investment does not tell us anything other than that the former can force out the latter. We, as a society, have a responsibility to choose the sort of agriculture we want and support it, not simply watch while big guys force out little guys and values we may not like in the long run supplant ones we would have wished to nurture had we thought of it.

We can spin out speculative scenarios like this about giant animals or the genetically engineered animals of the future until the cows come home, as it were. But enough has been said to illustrate the pitfalls of unreflectively embracing patenting of animals by those who are in a position to create such animals. Everything we have explored seems to support the current federal legislative thrust toward a moratorium on patenting. We need time, as a society, to think about and work through the sorts of issues I have raised and others that will come up, before we either embrace or reject animal patenting. I would, therefore, support such a five-year moratorium, pro vided that it included a mechanism and timetable for social discussion of these issues, aimed at generating a regulatory structure of the sort I have described at the end of the moratorium. We cannot afford to keep stalling this issue – the technology is growing exponentially, and if we do not act expeditiously, we will be dealing with a fait accompli, unshaped by national democratic, ethical, and prudential deliberation. Those companies who are at the forefront of biotechnology need ethical guidance from society, need to know the ground rules so that they can respect them and move ahead. As one biotechnologist said to me, with admirable candor: "Just tell us what rules to follow, and we will do it." But formulating these rules is a democratic responsibility, one that is unwise, immoral, and imprudent to shirk. The issues we have examined will not resolve themselves or go away, and, if left alone, they will proliferate.

CONCLUSION

Genetic engineering of animals, indeed all of biotechnology, is a tool, like all tools humans have deployed, from clay pots to nuclear reactors. In and of themselves, tools are neutral – pistols, surgical saws, and collection plates are not good and not bad. But tools are not just, or even primarily, in and of themselves. They are, as Heidegger says, ready at hand. They exist as tools in relation to the humans who deploy them, and they derive their "toolness," their telos, from the uses to which they are put. Unlike humans or animals, whose telos is in them, not in our uses of them, tools gain their natures from the set of real possibilities for their deployment. One can, I suppose, plow one's garden with a samurai sword, or crush garlic with an Uzi. But these are not part of the normal set of possibilities these instruments carry.

What in large part determines the normal set of possibilities for tools are the valuational contexts that surround them. A society whose values preclude any violence at all will, when encountering a cache of swords, either leave them untouched or develop its own set of possibilities for them, perhaps as garden or kitchen implements, but certainly not as weapons. Genetic engineering of animals stands to us and our society in something like that pristine relationship. There are those who would deploy it as a shortcut to profit, efficiency, and animal exploitation. If their values prevail, the technology will probably be bad for people and animals. On the other hand, its use can be informed by values of respect for animals, humans, nature, and risk and by notions of sustainability and awe at its potential. Which option develops is not something that will happen to us, but something we will do, be it by commission or forbearance. In this book, I have attempted to provide a basis for the knowledge and understanding that will empower us, collectively, to choose wisely.

APPENDIX

What is genetic engineering of animals, and how is it done?

All living things are a marvelous admixture of commonality and uniqueness. While all daisies express the features of "daisiness," and all pigs display "pigness," no two are exactly alike. The blueprint for both species' commonality and individuality is carried by the genes, which instruct and regulate the animals in how to develop, grow, and form throughout life. These genes are all sequences of DNA, an amazing molecule that has the ability to carry these instructions in all cells of the body and to self-replicate. Like the language of Morse code, which can carry the most complex messages using only two symbols, dots and dashes, DNA contains only four significatory components, and all genes are thus information-carrying sequences of these components.

Each animal, then, has a genetic program that directs it to develop into a pig, and into Porky, this pig. The genetic program for pigs or other living things undergoes changes through reproduction, when information from two individuals is combined to generate a new individual, and through artificial or natural selection, through which humans or nature determine which genetic programs will fit human or natural needs. When we breed dogs for certain traits, we choose to perpetuate certain genes and suppress others. When natural conditions favor one set of traits in an animal, say protective coloring versus coloring that flags the animal for predators, nature is choosing which genes will survive.

Genetic engineering is the process of taking new genes, never before incorporated into the program for the animal in question, and attempting to incorporate these genes not only into an individual, but into a multigenerational group who will carry the genes, and the traits they code for, and pass them on to subsequent generations. This allows for changing animals much more quickly than through breeding or evolution. The animals that carry the new gene or genes are called transgenic animals, and there are, at present, three methods for initiating the process of making transgenic animals. All involve inserting the foreign DNA into the animals' genetic program and then attempting to breed such animals consistently.

By far the most commonly used method is called *pronuclear injection* (see Figure 1). This method proceeds as follows: A female animal (most usually, for reasons of convenience and expense, a mouse) is hormonally treated so as to cause her to produce a large number of eggs and then mated with a fertile male. One- or two-celled embryos are flushed from the mother. At this point, several hundred copies of the gene to be inserted are injected, using a micropipette, into a portion of the embryo known as the male pronucleus, which contains the DNA contributed by the sperm to the embryo. (The male pronucleus is used 70 percent of the time in mice because it is usually easier to visualize than the female.) This is done with micromanipulators by someone visualizing the embryo through a microscope. When the process is successful (only 10 to 20 percent of the time), one or more than one copy of the gene inserts itself into one of the chromosomes (the part of the cell carrying the genes). Interestingly enough, where the gene inserts can affect how the gene later expresses itself. Ideally, as the young embryo grows and develops, each cell in the body, including germ cells, which will later pass genes to offspring, receives a copy of the new gene.

There are two major drawbacks to this method, although it works rather reliably. First, as just mentioned, is that the new genes are not expressed in a predictable way in the organism; how they express can vary with where and how many genes insert. Second, one can only introduce genes at a very early stage of embryo development.

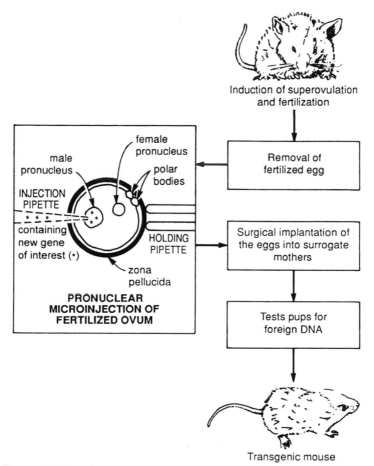

Figure 1. Development of a transgenic mouse. Source: *Research Resources Reporter,* January 1991.

The second method of introducing new DNA (genes) into an embryo is called *retroviral infection*. A retrovirus is a special sort of virus that can carry foreign DNA (actually an RNA version that is subsequently made into DNA) into a host. The retrovirus is modified in such a way that first of all it is not harmful and, secondly, it carries the desired new gene into chromosomes of the recipient embryo and infects all cells, including germ cells. This can be used at all stages of embryo development. Retrovirus infection thus can be used to make transgenic animals. (It can also be used, postembryonically, for gene therapy to treat genetic disease when it is used to infect the body cells containing the defective gene with non-defective genes. As mentioned earlier, such treatment does not affect the germ cells, so the cure does not pass on into the next generation.)

The third method of producing transgenic animals is through the use of embryonic stem cells or ES cells. These are cells taken from the blastocysts (an early stage of development) and kept alive in culture, where they grow and reproduce but do not differentiate. These cells can be taken out of culture and inserted into blastulas where they incorporate into the cell mass, including germ cells. Utilizing this concept, one can introduce foreign genes into ES cells while they are in culture, where they will grow and reproduce with the foreign gene incorporated into them. When the ES cells containing the foreign gene are reintroduced into blastocysts, one can produce the desired transgenic animal.

However one inserts the foreign gene, to get a transgenic line the foreign gene must have been incorporated into the germ cells (i.e., sperm and egg). This is called a *founder animal*. The founder animal is then bred to a nontransgenic mate. Statistically, 50 percent of the resulting offspring will carry the new gene in all their cells. (The other 50 percent will not have the gene at all.) One then breeds related animals, usually father-daughter or brother-sister, to produce a second generation of offspring, 25 percent of which, statistically, will be homozygous for the new gene. (That is, they will carry only the new gene and no others at the relevant part of the chromosome.) By breeding two such homozygous animals, one will then get 100 percent of offspring carrying the new gene. In this way, one has produced a strain of transgenic animals that can be predictably bred.

Notes

INTRODUCTION

1. B. E. Rollin, " 'The Frankenstein Thing': The Moral Impact of Genetic Engineering of Agricultural Animals on Society and Future Science."
2. D. F. Glut, *The Frankenstein Catalog*.
3. "A Vision of Ourselves," *Time*, 29 July 1985, 55.
4. U.S. Congress, Office of Technology Assessment, "New Developments in Biotechnology: Public Perceptions of Biotechnology."
5. B. E. Rollin, *The Unheeded Cry*.

1. "THERE ARE CERTAIN THINGS HUMANS WERE NOT MEANT TO DO"

1. Rollin, *The Unheeded Cry*, chap. 1.
2. Ibid., passim.
3. W. T. Keeton and J. L. Gould, *Biological Science*, p. 6.
4. S. Mader, *Biology*, p. 15.
5. J. Katz, "The Regulation of Human Experimentation in the United States – A Personal Odyssey."
6. C. Holden, "Universities Fight Animal Activists."
7. Michigan State *News*, 27 February 1989, p. 8.
8. Office of Technology Assessment, "New Developments in Biotechnology: Public Perceptions of Biotechnology."
9. National Science Foundation, Science Indicators 1980.
10. M. Crawford, "Religious Groups Join Animal Patent Battle."
11. Statement of 24 religious leaders against animal patenting.
12. Ibid.
13. D. Bonhoeffer, *Letters and Papers from Prison*.
14. J. Dewey, *A Common Faith*.
15. J. Rifkin, *Declaration of a Heretic*, p. 53.

16. Democritus, in J. R. Kirk and J. E. Raven, *The Presocratic Philosophers*, p. 422.
17. Aristotle, *Metaphysics*, bk. 1, chap. 3.
18. Preamble to the Constitution of the World Health Organization, in T. A. Mappes and J. S. Zembaty, *Biomedical Ethics*, p. 202ff.
19. O. Sacks, *Awakenings*.
20. A. O. Lovejoy, *The Great Chain of Being*, p. 59.
21. Rifkin, *Declaration of a Heretic*, p. 53.
22. Ibid.
23. Ibid.
24. B. H. Burma, "The Species Concept: A Semantic Review."
25. E. Mayr, "The Species Concept: Semantics Versus Semantics."
26. M. Ruse, *The Philosophy of Biology*, p. 127.
27. Mayr, "The Species Concept," p. 371.
28. Keeton and Gould, *Biological Science*, p. 883.
29. Mader, *Biology*, p. 745.
30. Professor D. Pettus, personal communication.
31. B. Rollin, *Natural and Conventional Meaning: An Examination of the Distinction*.
32. B. Rollin, "Nature, Convention, and Genre Theory."
33. R. J. Lifton, *The Nazi Doctors*.
34. Rollin, *The Unheeded Cry*, chaps. 5 and 6.
35. Sacks, *Awakenings*.
36. B. Rollin, "On the Nature of Illness."
37. K. Menninger, *Human Mind*.
38. I. Kant, *Foundations of the Metaphysics of Morals*.
39. H. Rolston, *Environmental Ethics* and *Philosophy Gone Wild*, passim.
40. Rolston, "Are Values in Nature Subjective or Objective," *Philosophy Gone Wild*, p. 114.
41. Ibid., p. 42.
42. Ibid., p. 44.
43. Rolston, *Environmental Ethics*, p. 112.
44. Ibid.
45. Ibid., p. 114.
46. Ibid., p. 116.
47. Ibid., p. 198.
48. Rolston, *Philosophy Gone Wild*, pp. 183, 212.
49. Ibid., p. 183.
50. E. Abbey, *Desert Solitaire*.
51. H. Rolston, "Duties to Endangered Species," in *Philosophy Gone Wild*, pp. 211 ff.
52. Ibid., p. 213.
53. Ibid., pp. 212–13.

54. Ibid., pp. 215–16.
55. Ibid., pp. 206 ff.
56. Rolston, *Environmental Ethics*, p. 59.
57. P. Ramsey, *Fabricated Man*, p. 94.
58. Ibid., p. 96.

2. RAMPAGING MONSTERS

1. R. Bradbury, "A Sound of Thunder."
2. P. Ehrenfeld, *The Arrogance of Humanism*.
3. S. Krimsky, *Biotechnics and Society*, p. 100.
4. J. Katz, "The Regulation of Human Experimentation in the United States – A Personal Odyssey."
5. M. Crichton, *Jurassic Park*.
6. Ibid., pp. 76–77.
7. B. Rollin, *The Unheeded Cry*, chap. 1.
8. D. Hume, *A Treatise of Human Nature*, p. 272.
9. M. A. Liebert, "From the Publisher," *Genetic Engineering News*, 1 March 1992, p. 4.
10. Ibid.
11. C. Potera, "Will 'Jurassic Park,' the Movie, Create a PR Problem for Biotechnology?," *Genetic Engineering News*, 1 March 1992, p. 22.
12. Ibid., p. 23.
13. Ibid.
14. N. Wiener, *The Human Use of Human Beings*.
15. N. Rescher, *Risk*, pp. 26–32.
16. B. Fischhoff et al., *Acceptable Risk*, pp. 79 ff.
17. T. Hoban and P. Kendall, *Consumer Attitudes about the Use of Biotechnology in Agriculture and Food Production*.
18. B. E. Rollin, "Nature, Convention, and the Medical Approach to the Dying Elderly."
19. B. E. Rollin, "Some Ethical Concerns in Animal Research: Where Do We Go Next?"
20. P. K. Feyerabend, *Science in a Free Society*.
21. D. E. Albrecht, "Public Perceptions of Agricultural Biotechnology: An Overview of Research to Date," p. 62.
22. Ibid.
23. Hoban and Kendall, *Consumer Attitudes about the Use of Biotechnology in Agriculture and Food Production*.
24. Fort Collins *Coloradoan*, 26 May 1992, 1.
25. Ibid.
26. B. E. Rollin, *Animal Rights and Human Morality*, part 4.

27. Personal communication.
28. J. M. Leonard et al., "Development of Disease and Virus Recovery in Transgenic Mice Containing Proviral DNA."
29. D. E. Mosier et al., "Transfer of a Functional Human Immune System to Mice with Severe Combined Immunodeficiency."
30. P. Lusso et al., "Expanded HIV-1 Cellular Tropism by Phenotypic Mixing with Murine Endogenous Retroviruses."
31. J. M. Tiedje et al., "The Planned Introduction of Genetically Engineered Organisms: Ecological Considerations and Recommendations."
32. *Fisheries*, vol. 15, no. 1, January–February 1990.
33. Tiedje et al., "The Planned Introduction of Genetically Engineered Organisms," p. 302.
34. Ibid.
35. Ibid., p. 305.
36. Ibid., p. 311.
37. *Fisheries*, pp. 2, 12, 21.
38. Ibid., p. 9.
39. Ibid.

3. THE PLIGHT OF THE CREATURE

1. Dr. Randall Lockwood, personal communication.
2. See B. E. Rollin, *Animal Rights and Human Morality*, second edition, part 2.
3. *Animal Legal Defense Fund v. The Department of Environment Conservation of the State of New York*.
4. B. E. Rollin, *The Unheeded Cry*, chap. 7.
5. S. Suther, "In Search of Common Ground," and personal communication.
6. *Parents Poll on Animal Rights, Attractiveness, Television, and Abortion*.
7. Dr. Leo Bustad, personal communication.
8. R. Harrison, *Animal Machines*.
9. *Parents Poll on Animal Rights, Attractiveness, Television, and Abortion*.
10. "Swedish Animals Get a Bill of Rights," *New York Times*, 25 October 1988.
11. Rollin, *Animal Rights and Human Morality*, part 2.
12. B. E. Rollin, "Federal Laws and Policies Governing Animal Research: Their History, Nature, and Adequacy."
13. B. E. Rollin, *The Unheeded Cry*.
14. "Swedish Animals Get a Bill of Rights."
15. B. E. Rollin and M. L. Kesel (eds.), *The Experimental Animal in Biomedical Research*, vol. 2.

16. "Swedish Animals Get a Bill of Rights."
17. J. Mench, "The Welfare of Poultry in Modern Production Systems."
18. R. Warner, quoted in S. Suther, "In Search of Common Ground."
19. D. Gärtner et al., "Stress Response of Rats to Handling and Experimental Procedures."
20. V. Riley, "Mouse Mammary Tumors: Alteration of Incidence As Apparent Function of Stress."
21. Rollin, *Animal Rights and Human Morality*, part 3.
22. V. Pursel et al., "Genetic Engineering of Livestock"; R. J. Wall and G. E. Seidel, "Transgenic Farm Animals – A Critical Analysis."
23. Ibid.
24. Ibid.
25. M. Fox, *Superpigs and Wondercorn*, p. 117.
26. Pursel et al., "Genetic Engineering of Livestock."
27. D. B. Lewis et al., "Osteoporosis Induced in Mice by Overproduction of Interleukin-4."
28. Gordon Niswender, personal communication.
29. J. D. McNeish et al., "Legless, a Novel Mutation Found in PHT1-1 Transgenic Mice."
31. Ibid.
32. Ibid.
33. Fox, *Superpigs and Wondercorn*, p. 103.
34. L. Henninghausen et al., "Transgenic Animals – Production of Foreign Proteins in Milk."
35. Wall and Seidel, "Transgenic Farm Animals."
36. G. Seidel, "Transgenic Animals."
37. Ibid.
38. Wall and Seidel, "Transgenic Farm Animals."
39. B. E. Rollin, "Transgenic Animals: Science and Ethics."
40. D. Hanahan, "Transgenic Mice As Probes into Complex Systems."
41. R. Jaenisch, "Transgenic Animals."
42. Hanahan, "Transgenic Mice As Probes"; Jaenisch, "Transgenic Animals."
43. B. E. Rollin, "Transgenic Animals: Science and Ethics."
44. Rollin, *The Unheeded Cry*; M. Pernick, *A Calculus of Suffering*.
45. E. M. Karson, "Principles of Gene Transfer and the Treatment of Disease."
46. J. W. Gordon, "State of the Art: Transgenic Animals."
47. Ibid.
48. R. Kolberg, "Gene Therapy: Animal Models Point the Way to Human Clinical Trials."
49. M. Zoler, "Transgenic Animal Model for Human Inflammatory Disease."
50. Karson, "Principles of Gene Transfer."

51. C. R. Scriver et al., *The Metabolic Basis of Inherited Disease*, vols. 1 and 2.
52. M. Hooper et al., "HPRT-Deficient (Lesch-Nyhan) Mouse Embryos Derived from Germline Colonization by Cultured Cells"; M. R. Kuehn et al., "A Potential Model for Lesch-Nyhan Syndrome through Introduction of HPRT Mutations into Mice."
53. W. N. Kelley and J. B. Wyngaarden, "Clinical Syndromes Associated with Hypoxanthine-Guanine Phosphororibosyltransferase Deficiency."
54. Ibid.
55. J. T. Stout and C. T. Caskey, "Hypoxanthine Phosphororibosyltransferase Deficiency: The Lesch-Nyhan Syndrome and Gouty Arthritis."
56. Kelley and Wyngaarden, "Clinical Syndromes."
57. E. M. Boyd et al., "The Chronic Oral Toxicity of Caffeine."
58. Hooper et al., "HPRT-Deficient (Lesch-Nyhan) Mouse Embryos"; Kuehn et al., "A Potential Model for Lesch-Nyhan Syndrome."
59. Stout and Caskey, "Hypoxanathine Phosphororibosyltransferase Deficiency."
60. Fox, *Superpigs and Wondercorn*, p. 104.
61. Ibid.

References

"A Vision of Ourselves," *Time*, 29 July 1985, 5.

Abbey, E., *Desert Solitaire: A Season in the Wilderness* (New York: McGraw-Hill, 1968).

Albrecht, D. E., "Public Perceptions of Agricultural Biotechnology: An Overview of Research to Date," in *Ethics and Patenting of Transgenic Organisms: NABC Occasional Papers* #1 (Ithaca, New York: NABC, 1992).

Animal Legal Defense Fund v. *The Department of Environment Conservation of the State of New York*. Index # 6670/85.

Aristotle, *Metaphysics*.

Bonhoeffer, D., *Letters and Papers from Prison* (London: SCM Press, 1967).

Boyd, E. M. et al., "The Chronic Oral Toxicity of Caffeine," *Canadian Journal of Physiology and Pharmacology* 43 (1965), 94 ff.

Bradbury, R., *R Is for Rocket* (New York: Bantam, 1965).

Burma, B. H., "The Species Concept: A Semantic Review," *Evolution*, vol. 3 (1949), 369–70.

Crawford, M., "Religious Groups Join Animal Patent Battle," *Science* 237 (1987), 480–1.

Crichton, M., *Jurassic Park* (New York: Knopf, 1990).

Dewey, J., *A Common Faith* (New Haven: Yale University Press, 1934).

Ehrenfeld, D., *The Arrogance of Humanism* (New York: Oxford University Press, 1978).

Evans, W. J. and A. Hollaender (eds.), *Genetic Engineering of Animals: An Agricultural Perspective* (New York: Plenum Press, 1986).

Feyerabend, P. K., *Science in a Free Society* (London: NLB, 1978).

First, N., and F. P. Haseltine (eds.), *Transgenic Animals* (Boston: Butterworth-Heinemann, 1991).

Fischhoff, B., S. Lichtenstein, P. Slovic, S. Derby, R. Keeney, *Acceptable Risk* (Cambridge: Cambridge University Press, 1981).

Fox, M., *Superpigs and Wondercorn* (New York: Lyons and Burford, 1992).

References

Gärtner, D. et al., "Stress Response of Rats to Handling and Experimental Procedures," *Laboratory Animals* 14 (1980) 267–74.

"Genetically Engineered Foods Won't Get Special FDA Treatment," Fort Collins *Coloradoan*, 26 May 1992, 1.

Glut, D. F., *The Frankenstein Catalog: Being a Comprehensive History of Novels, Translations, Adaptations, Stories, Critical Works, Popular Articles, Series, Fumetti, Verse, Stage Plays, Films, Cartoons, Puppetry, Radio and Television Programs, Comics, Satire and Humor, Spoken and Musical Recordings, Tapes and Sheet Music Featuring Frankenstein's Monster and/or Descended from Mary Shelley's Novel* (Jefferson, North Carolina: McFarland, 1984).

Gordon, J. W., "State of the Art: Transgenic Animals," *ILAR News*, vol. 30, no. 3 (1988), 8–17.

Hallerman, E. M., and A. R. Kapuscinski, "Transgenic Fish and Public Policy: Anticipating Environmental Impacts of Transgenic Fish," *Fisheries*, vol. 15, no. 1 (1990), 2–12.

"Transgenic Fish and Public Policy: Patenting Transgenic Fish," *Fisheries*, vol. 15, no. 1 (1990), 21–5.

"Transgenic Fish and Public Policy: Regulatory Concerns," *Fisheries*, vol. 15, no. 1 (1990), 12–21.

Hanahan, D., "Transgenic Mice as Probes into Complex Systems," *Science* 246 (1989), 1265–75.

Harrison, R., *Animal Machines* (London: Vincent Stuart, 1964).

Henninghausen, L. et al., "Transgenic Animals – Production of Foreign Proteins in Milk," *Current Opinion in Biotechnology*, vol. 1 (1990), 74–8.

Hoban, T., and P. Kendall, *Consumer Attitudes About the Use of Biotechnology in Agriculture and Food Production*, Interim Report, Colorado State University and North Carolina State University, July 1992.

Holden, C., "Universities Fight Animal Activists," *Science* 243 (1989), 17.

Hooper, M. et al., "HPRT-deficient (Lesch-Nyhan) Mouse Embryos Derived from Germline Colonization by Cultured Cells," *Nature* 326 (1987), 292–5.

Hume, D., *A Treatise of Human Nature* (Oxford: Oxford University Press, 1960).

Isola, L. M., and J. W. Gordon, "Transgenic Animals: A New Era in Developmental Biology and Medicine," in N. First and F. P. Haseltine (eds.), *Transgenic Animals* (Boston: Butterworth- Heinemann, 1991).

Jaenisch, R., "Transgenic Animals," *Science* 240 (1988), 1468–73.

Kant,I., *Foundations of the Metaphysics of Morals*.

Karson, E. M., "Principles of Gene Transfer and the Treatment of Disease," in N. First and F. P. Haseltine (eds.), *Transgenic Animals* (Boston: Butterworth-Heinemann, 1991), chap. 16.

References

Katz, J., "The Regulation of Human Experimentation in the United States – A Personal Odyssey," *IRB* 9, no. 1 (1987), 1–6.

Keeton, W. T., and J. L. Gould, *Biological Science* (New York: W. W. Norton, 1986).

Kelley, W. N., and J. B. Wyngaarden, "Clinical Syndromes Associated with Hypoxanthine-Guanine Phosphororibosyltransferase Deficiency," in J. B. Stanbury et al. (eds.), *The Metabolic Basis of Inherited Disease*, fifth edition (New York: McGraw- Hill, 1983), chap. 51.

Kirk, J. R., and J. E. Raven, *The Presocratic Philosophers* (Cambridge: Cambridge University Press, 1957).

Kohlberg, R., "Gene Therapy: Animal Models Point the Way to Human Clinical Trials," *Science* 256 (1992), 772–3.

Krimsky, S., *Biotechnics and Society: The Rise of Industrial Genetics* (New York: Praeger, 1991).

Kuehn, M. R. et al., "A Potential Model for Lesch-Nyhan Syndrome through Introduction of HPRT Mutations into Mice," *Nature* 326 (1987), 295–8.

Leonard, J. M. et al., "Development of Disease and Virus Recovery in Transgenic Mice Containing HIV Proviral DNA," *Science* 242 (1988), 1665.

Lewis, D. B. et al., "Osteoporosis Induced in Mice by Overproduction of Interleukin–4," *Proceedings of the National Academy of Sciences* 90, no. 24 (1993), 11618–22.

Liebert, M. A., "From the Publisher," *Genetic Engineering News*, 1 March 1992, 4.

Lifton, R. J., *The Nazi Doctors: Medical Killing and the Psychology of Genocide* (New York: Basic Books, 1986).

Lovejoy, A. O., *The Great Chain of Being: A Study of the History of an Idea* (Cambridge, Mass.: Harvard University Press, 1936).

Lusso, P. et al., "Expanded HIV–1 Cellular Tropism by Phenotypic Mixing with Murine Endogenous Retroviruses," *Science* 247 (1990), 848–51.

Mader, S., *Biology: Evolution, Diversity, and the Environment* (Dubuque, Iowa: W. E. Brown, 1987).

Mappes, T. A., and J. S. Zembaty, *Biomedical Ethics* (New York: McGraw-Hill, 1981).

Martin, J. (ed.), *High Technology and Animal Welfare* (Edmonton: University of Alberta, 1991).

Marx, J., "Concerns Raised about Mouse Models for AIDS," *Science* 247 (1990), 809.

Mayr, E., "The Species Concept: Semantics versus Semantics," *Evolution*, vol. 3 (1949), 371–2.

McNeish, J. D. et al., "Legless, a Novel Mutation Found in PHT1–1 Transgenic Mice," *Science* 241 (1988), 837–9.

References

Mench, J., "The Welfare of Poultry in Modern Production Systems," *CRC Critical Reviews in Poultry Biology*, vol. 4, 1992, 107–128.

Menninger, K., *Human Mind* (New York: Knopf, 1930).

Mosier, D. E. et al., "Transfer of a Functional Human Immune System to Mice with Severe Combined Immunodeficiency," *Nature*, vol. 335 (1988), 256–9.

National Science Foundation, Science Indicators 1980.

Office of Technology Assessment, "New Developments in Biotechnology: Public Perceptions of Biotechnology" (Washington, D.C.: Office of Technology Assessment, 1987).

Parents Magazine, *Parents Poll on Animal Rights, Attractiveness, Television, and Abortion* (New York: Kane and Parsons Associates, 1989).

Pernick, M., *A Calculus of Suffering: Pain, Professionalism and Anesthesia in Nineteenth-Century America* (New York: Columbia University Press, 1985).

Potera, C., "Will 'Jurassic Park,' the Movie, Create a PR Problem for Biotechnology?," *Genetic Engineering News*, 1 March 1992, 22–3.

President's Commission for the Study of Ethical Problems in Medicine and Biomedical and Behavioral Research, *Splicing Life: The Social and Ethical Issues of Genetic Engineering with Human Beings* (Washington, D.C.: U.S. Government Printing Office, 1983).

Pursel, V. et al., "Genetic Engineering of Livestock," *Science* 244 (1989), 1281–8.

Ramsey, P., *Fabricated Man: The Ethics of Genetic Control* (New Haven: Yale University Press, 1970).

Rescher, N., *Risk: A Philosophical Introduction to the Theory of Risk Evaluation and Management* (Lanham, Md.: University Press of America, 1983).

Rifkin, J., *Algeny* (New York: Viking Press, 1983).

Declaration of a Heretic (Boston: Routledge and Kegan Paul, 1985).

Riley, V., "Mouse Mammary Tumors: Alteration of Incidence As Apparent Function of Stress," *Science* 189 (1975), 465.

Rollin, B. E., *Animal Rights and Human Morality*, second edition (Buffalo, N.Y.: Prometheus Books, 1992).

"Environmental Ethics and International Justice," in S. Luper-Foy (ed.), *Problems of International Justice* (Boulder, Colo.: Westview Press, 1988).

"Ethics and Research Animals – Theory and Practice," in B. E. Rollin and M. L. Kesel (eds.), *The Experimental Animal in Biomedical Research*, vol. 1 (Boca Raton, Fla.: CRC Press, 1990), chap. 2.

"Federal Laws and Policies Governing Animal Research: Their History, Nature, and Adequacy," in J. M. Humber and R. F. Almeder, *Biomedical Ethics Reviews: 1990* (Clifton, N.J.: Humana, 1991).

"The Frankenstein Thing," in J. W. Evans and A. Hollaender (eds.),

References

Genetic Engineering of Animals: An Agricultural Perspective (New York: Plenum, 1986), 285–98.

Natural and Conventional Meaning: An Examination of the Distinction (The Hague: Mouton and Co., 1976).

"Nature, Convention, and Genre Theory," *Poetics*, vol. 10 (1981), 127–43.

"Nature, Convention, and the Medical Approach to the Dying Elderly," in M. Tallmer et al. (eds.), *The Life-Threatened Elderly* (New York: Columbia University Press, 1984).

"On the Nature of Illness," *Man and Medicine* 4, no. 3 (1979), 157–72.

"Some Ethical Concerns in Animal Research: Where Do We Go Next?" in R. M. Baird and S. E. Rosenbaum, *Animal Experimentation: The Moral Issues* (Buffalo, N.Y.: Prometheus Books, 1991), 103–14.

"Transgenic Animals: Science and Ethics," in B. E. Rollin and M. L. Kesel (eds.), *The Experimental Animal in Biomedical Research*, vol. 2 (Boca Raton, Fla.: CRC Press, 1993).

The Unheeded Cry: Animal Consciousness, Animal Pain, and Science (Oxford: Oxford University Press, 1989).

Rollin, B. E. and M. L. Kesel (eds.), *The Experimental Animal in Biomedical Research*, vol. 2 (Boca Raton, Fla.: CRC Press, 1995).

Rolston, H., "Duties to Endangered Species," *Bioscience* 35 (1985), 718–26.

Environmental Ethics: Duties to and Values in the Natural World (Philadelphia, Pa.: Temple University Press, 1988).

Philosophy Gone Wild (Buffalo, N.Y.: Prometheus Books, 1989).

Ruse, M., *The Philosophy of Biology* (London: Hutchinson University Library, 1973).

Sacks, O., *Awakenings* (New York: Vintage, 1976).

Scriver, C. R. et al. (eds.), *The Metabolic Basis of Inherited Disease*, vols. 1 and 2, sixth edition (New York: McGraw-Hill, 1989).

Seidel, G. E., Jr., "Biotechnology in Animal Agriculture," in J. F. Macdonald (ed.). *National Agricultural Biotechnology Council Report 3: Biological, Social and Institutional Concerns* (Ithaca, N.Y.: NABC, 1991), 1–12.

"Transgenic Animals," in *Embryo Transfer*, vol. 7 (1992), 1–5.

Stanbury, J. B. et al., *The Metabolic Basis of Inherited Disease*, fifth edition (New York: McGraw-Hill, 1983).

Statement of 24 religious leaders against animal patenting, 1987. (Distributed by the Foundation for Economic Trends, no date or place given.)

Sterling, J., "FDA Approves Use of Bovine Somatotropin to Increase Milk Production in Dairy Cows," *Genetic Engineering News*, November 15, 1993, 1 ff.

Stout, J. T., and C. T. Caskey, "Hypoxanthine Phosphororibosyltransferase Deficiency: The Lesch-Nyhan Syndrome and Gouty Arthritis," in C. R. Scriver et al. (eds.), *The Metabolic Basis of Inherited Disease*, vol. 1 (New York: McGraw-Hill, 1989), chap. 38.

Suther, S., "In Search of Common Ground," *Beef Today*, March 1993.

"Swedish Animals Get a Bill of Rights," *New York Times*, 25 October 1988.

Tiedje, J. M. et al., "The Planned Introduction of Genetically Engineered Organisms: Ecological Considerations and Recommendations," *Ecology*, vol. 70, no. 2 (1989), 297–315.

"Transgenic Mouse Facility," *Research Resources Reporter*, vol. 15, no. 1 (1991), 13–14.

U.S. Congress, Office of Technology Assessment, *New Developments in Biotechnology: Patenting Life – Special Report* (Washington, D.C.: U.S. Government Printing Office, 1989).

U.S. Department of Health and Human Services, Public Health Service, Centers for Disease Control, and National Institutes of Health, *Biosafety in Microbiological and Biomedical Laboratories* (Washington, D.C.: U.S. Government Printing Office, 1988).

Wall, R. J. and G. E. Seidel, "Transgenic Farm Animals – A Critical Analysis," *Theriogenology* 38 (1992).

West, P., "Director Addresses Health Research," *Michigan State News* 27 February 1989, 8.

Wiener, N., *The Human Use of Human Beings: Cybernetics and Society* (Boston: Houghton Mifflin, 1950).

Zoler, M., "Transgenic Animal Model for Human Inflammatory Disease," *Research Resources Reporter*, vol. 16, no. 4 (1992), 1–4.

Index

Index

Civil Rights Act, 145–6
Civil Rights movement, 145–6
Clark, Stephen, 151
classification, 41, 42; of species, 37
closed systems, 74
Colorado State University Rodeo Club, 144–5
commercial purposes: genetic engineering for, 183–4, 185, 190–1, 208
commercial research: lack of regulation in, 122
common sense, 26, 37, 98; in animal welfare, 155, 168, 169; biotechnology in, 97–8; and genetic engineering of animals, 84, 104; and potential dangers of genetic engineering, 75; and risk assessment, 100; science and, 14, 15, 16, 73, 74–5
computers, 78
conceptual schemes, 41–2
confinement agriculture, 153, 157, 161, 171, 178, 186, 187, 190, 210; animal welfare in, 179, 180–1; laws regarding, 149; rejection of, 208
Congress, 69, 81, 148, 149
consciousness: agnosticism regarding, 14; eliminating, in transgenic animals, 205–6; see also sentience
consequential wrongness of genetic engineering, 40, 66, 175
conservation ethic, 36
conservation of welfare; see principle of conservation of welfare
convention: and nature, in genetic engineering, 40–7
corn, 110
corn blight, 110
corporations: and animal patenting, 210–11
cosmetics testing: animal use in, 140, 149, 166, 177
Council on Competitiveness, 99–100
creature (the): moral obligations to, 168–9; plight of, 137–214
Crichton, Michael, 72–3, 75, 76–7, 86; *Jurassic Park*, 72–3, 118–19, 122
cruelty to animals, 147–8, 151, 152, 158, 159, 164, 165, 177; moving beyond, in new ethic for animals, 154–68; see also anticruelty ethic
Curtis, Stanley, 178, 187
cysteine, 191–2
cystic fibrosis, 65

dairy farmers, 101, 102, 134
dangers, possible (genetic engineering), 71–7, 108–11; disease models, 115–17; environmental, 117–33; issue of, 67–9; in military applications, 133–4; narrowing of gene pool, 111–13; selecting for pathogens, 114–15; see also risk assessment; risk-benefit analysis
Darwin, Charles, 23, 24
decerebration, 193, 205
decision making, 82–4; by experts, 92; regarding genetic engineering, 102, 132; mutual, 99; public participation in, 98, 105; rational, 82–3, 94–5; see also social decision making
democracy: deference to experts in, 91; educability of public in, 89–90; and risk sharing, 79
democratic discussion: values in, 46
democratic regulatory mechanism: for genetic engineering of animals, 207–8, 213; see also regulation; regulatory structure
democratic risk assessment, 92–9
Democritus, 25, 40
Department of Defense, 148–9
desacralization, 24–5, 66
Descartes, René, 14, 42; *Meditations*, 19
desertification, 132–3
developmental biology, 196–7
Dewey, John, 24
dialogue, 98; ethical, 145; on risks of genetic engineering, 88, 89–90, 99, 107
diethylstilbestrol, 111
dilemmas, 9–10
disease, 3, 25; animal parts in treatment of, 63; reductionist approach to, 26–7; values in, 43–6; see also genetic defects/diseases
disease models, genetically engineered, 115–17, 182, 198–9, 200–6, 208
disease research, 74
disease resistance, 114, 170, 179, 191
disenfranchised groups/minorities, 146, 156
diversity (human): loss of, 27
DNA, 96, 194, 196, 215; introducing into genes, 216, 217–18
DNA sequencing, 42, 110–11
dogs, 151, 163, 186–7, 215; genetic diseases in, 109–10, 170
dualism, 10–11; of nature and convention, 40–7

233

Index

Index